Does Aging Stop?

Does Aging Stop?

LAURENCE D. MUELLER,
CASANDRA L. RAUSER, AND
MICHAEL R. ROSE

OXFORD
UNIVERSITY PRESS

OXFORD
UNIVERSITY PRESS

Oxford University Press, Inc., publishes works that further
Oxford University's objective of excellence
in research, scholarship, and education.

Oxford   New York
Auckland   Cape Town   Dar es Salaam   Hong Kong   Karachi
Kuala Lumpur   Madrid   Melbourne   Mexico City   Nairobi
New Delhi   Shanghai   Taipei   Toronto

With offices in
Argentina   Austria   Brazil   Chile   Czech Republic   France   Greece
Guatemala   Hungary   Italy   Japan   Poland   Portugal   Singapore
South Korea   Switzerland   Thailand   Turkey   Ukraine   Vietnam

Published by Oxford University Press, Inc.
198 Madison Avenue, New York, New York 10016

www.oup.com

Oxford is a registered trademark of Oxford University Press

Library of Congress Cataloging-in-Puplication Data
Mueller, Laurence D., 1951–
Does aging stop?/ Laurence D. Mueller, Casandra L. Rauser, Michael R. Rose.
p. ; cm.
Includes bibliographical references and index.
ISBN 978-0-19-975422-9 (hardcover : alk. paper)
1. Aging.   2. Human evolution.   3. Longevity.   4. Mortality.   I. Rauser,
Casandra L., 1974–   II. Rose, Michael R. (Michael Robertson), 1955–   III. Title.
[DNLM: 1. Aging—genetics.   2. Aging—physiology.   3. Demography.
4. Evolution.   5. Longevity.   WT 1047]
QP86.M84 2011
613.2—dc22   2010038541

To our teachers, mentors, and inspirations: Francisco Ayala, James F. Crow, Marc Feldman, Joseph Graves, the late William Hamilton, Rudolf Harmsen, Ronald Rutowski, Robert Sikes, the late John Maynard Smith, and the late Sewall Wright

The idea that aging is a physiological process of progressive debilitation and inevitable death, unless accident or infection kill first, has held sway in academic biology since the time of Aristotle. That the passing of years should steadily foster decrepitude is as intuitively natural for most biologists as the sun revolving about the Earth was for most astronomers before Copernicus.

In this book, we attempt to instigate a revolutionary event for the science of aging. We claim that aging stops at the level of the individual—and aim to explain why. We open a door that leads to a new approach to the scientific puzzles of aging. And we offer new practical possibilities for intervening in the human aging process.

Some of our colleagues are convinced that we are wrong. Once upon a time—in fact, just 20 years ago—we would have agreed with them. We started out as determinedly "Aristotelian" as almost any other scientists studying the problem of aging. We, too, thought it proceeded without remit to the point of sterility and death, and that assumption was incorporated into the very definition of aging that we offered in our earlier book, *Evolutionary Biology of Aging* (Rose 1991). For decades, we were convinced that our work, too, showed that aging was as inevitable as sunrise on a clear day.

But our contented conformity with prevailing gerontological assumptions was challenged by amazing results published in 1992 by the laboratories of two of our longstanding colleagues: Jim Carey and Jim Curtsinger. Curtsinger et al. (1992), in particular, were quite direct in pointing out that their work posed a major challenge to the *limited life span paradigm* that follows naturally from the Aristotelian assumption of unremitting aging. While we argued against their findings in print at first, their diligent work following up on their initial breakthrough progressively undermined our objections.

The problem became, at least for us, what to make of the basic demographic fact that Carey and Curtsinger had established: mortality rates eventually stop increasing at very late ages in at least some well-defined laboratory cohorts. That is, they established that demographic aging stops, at the level of aggregate mortality, as a brute empirical fact. And we didn't know how that was possible.

Around 1994, we started to get a glimmer of an explanation for this perplexing fact, and that led us to embark on the program of research that we describe in detail in this volume. We have used both formal theory and massive experimentation in our quest to understand the cessation of aging. We feel that we can now provide the outlines of our solution to this problem to the wider scientific community, beyond the narrow confines of the experimental evolutionists, theoretical demographers, and experimental gerontologists who have been our chief audience up to this point in time. In presenting a well-rounded account of our work here, we have penetrated deeper than we have in any of our previous publications, finding new areas of application and new scientific projects for future research.

We do not offer here a definitive or concluding statement of the new post-Aristotelian gerontology. Instead, this book is intended to mark a new beginning for the science of aging. Replacing the presumption of endless deterioration with the paradoxes that are produced by its cessation has led us to a materially different view of what we personally have been doing for a third of a century in our research on aging. Our hope is that the aging field will now reconsider the assumptions upon which it has been based for more than two millennia. Perhaps the sun doesn't rise above a recumbent Earth on brightly lit mornings, after all. And perhaps aging stops, eventually.

Irvine, California
July 2010

ACKNOWLEDGMENTS

This book relies on data that were collected primarily by hundreds of University of California, Irvine, undergraduate research students in our laboratories over the 15 years that we have been working on the cessation of aging. We want to thank them all for their hard work, but especially Yasmine Abdel-Aal, Gabriel Covarrubias, Michelle Cung, Sabrina Gunion, Justin Hong, Kathy Pham, Jonathan Shieh, Christine Suen, Rena Thakar, John Tierney, Puya Yazdi, Suhail Beguwala, Diana Moise, Tuan Ly, Avi Mamidi, Ali Jazayeri, Christopher Hammerle, Erin Gass, Michael Simison, Kandarp Shah, Kristy Doyal, Ronald Ly, Simon Lee, Shahin Yousef, Rupesh Kalthia, Francis Rocha, Melinda Caldejon, Vannina Suarez, Eric Morgan, Sean Sepulveda, Irvin Bussel, and Ian Frazier.

Furthermore, our proximal colleagues Curtis Adams, Greg Benford, Molly Burke, Adam Chippindale, Mark Drapeau, Joseph Graves, Amitabh Joshi, Margarida Matos, Ted Nusbaum, and Parvin Shahrestani have made major and indispensable contributions to varied aspects of our research in this area. The book benefited from the numerous comments of various colleagues, although we have not always followed their advice. In particular, we thank Molly Burke, Caleb Finch, Steve Frank, George Martin, Margarida Matos, Scott Pletcher, Parvin Shahrestani, and David Steinsaltz for their reviews of draft material.

We would also like to thank our distal colleagues James Carey, James Curtsinger, Kenneth Wachter, and Brian Charlesworth for their inspirational and useful publications in this area. However much we may have agreed or disagreed over the years, their work has been of great importance for the development of our thinking, our theoretical work, and our experimental strategies.

Over the many years of work on this project, we have received financial support from a variety of agencies and institutions, including the National Institutes of Health, Sigma Xi, the National Science Foundation, the U.S.

Department of Education, the American Association of University Women, and the University of California.

Tisse Takagi has provided unstinting encouragement and understanding throughout the writing, reviewing, and production process of this book. We have been fortunate to have her thoroughly professional support.

Last, but certainly not least, we appreciate our long-suffering departmental colleagues and family members for their patience, support, and encouragement, especially Blanca Cervantes, Carol Krieks, and Brandon McDonald.

CONTENTS

FIGURES

TABLES

Does Aging Stop?

Introduction: A Third Phase of Life

There is a third phase of life, following development and aging, that we call late life. Late life requires different evolutionary, demographic, and physiological principles from those that characterize the first two phases of life.

THE FIRST TWO PHASES OF LIFE

Life histories have traditionally been organized into two phases: *development* and *adulthood*. The dividing line between these phases is the onset of reproductive maturity, an event that is usually easily discerned.

A major qualification to this categorization arises in organisms that reproduce by symmetrical fission. Among the organisms that reproduce this way are unicellular species, such as many bacteria, some protozoa, and some algae. Some species of both multicellular plants and multicellular animals reproduce by approximately symmetrical fission as well. Fissile reproduction occurs in free-living coelenterates, for example, particularly sea anemones and some *Hydra* species. It is also common among clonally spreading plants, such as grasses, some herbs, and even trees. Among species that reproduce by symmetrical fission, there is no adult phase of appreciable duration. Juveniles effectively reproduce two more juveniles, if they survive to reach reproductive competence. Simply put, there is no true adulthood.

Most multicellular species, however, do attain an adulthood during which they reproduce with a soma left behind, a structure that does not itself join the next generation as an immature organism. Such asymmetrical reproduction is empirically associated with the occurrence of *aging*, in which the soma progressively deteriorates with time, even under good conditions, regardless of excellent provisioning and protection from predators or obvious contagious disease. That is, adulthood is empirically associated with endogenous deterioration under good conditions. When adult mortality rates in species with asymmetrical reproduction are studied carefully under protected conditions, particularly in laboratories, their mortality rates typically accelerate rapidly after the onset of reproduction.

There is a paradoxical element to this pattern. Multicellular organisms typically undergo a complex process of cell proliferation and differentiation during development, before the onset of adulthood. Then, shortly after the organism is fully developed, with all of its specialized tissues available for use, it proceeds to deteriorate. This deterioration is slow at first, but it progressively accelerates over a long period of time, reducing the likelihood of continued survival to much lower levels than are observed at the start of adulthood. It is seemingly a contradiction of adaptive evolution, given that the life-cycle of a complex organism evolves to proceed successfully through a complex process of development, which often takes place under conditions of uncertain access to energy as well as threats of predation or mechanical destruction. Yet the reproductive adult that has survived this process of development soon begins a process of pervasive deterioration even under the most benign conditions that experimenters can contrive.

Irrespective of how these two phases of life are explained, they are striking in their distinctness. As a result, two very different scientific fields have focused on the two widely recognized phases of life: developmental biology and gerontology. The former discipline, developmental biology, was one of the most successful twentieth-century biological disciplines. The latter discipline, gerontology, hasn't been quite as successful, though perhaps primarily for institutional and historical reasons more than scientific deficiencies, as we will discuss.

Within these two disciplines, a wide spectrum of research strategies have been used, from biochemistry to molecular biology to cell biology to organismal physiology to population genetics to evolutionary theory. However, it is fairly natural to separate this research into two capacious bins: (i) molecular, cellular, or reductionist research versus (ii) comparative, demographic, or evolutionary research. We discuss each in turn.

THE SCIENTIFIC STUDY OF THE FIRST TWO PHASES OF LIFE: MOLECULAR AND CELL BIOLOGY

The study of the molecular and cell biology of development has been extremely successful. Piece by piece, the machinery by which adult organisms are produced has been detailed, deconstructed, and manipulated. This is such a well-attested success story in biology that we will devote no further attention to it here.

By contrast, it has been extremely difficult to work out the molecular and cell biology of aging, leaving aside the mere documentation of its numerous changes. The most interesting result has been that the lifespan can be "stretched" in organisms with artificially lower metabolic or reproductive rates (Weindruch and Walford 1988; Finch 1990), although this result had already been demonstrated as early as 1916 and 1917 (Loeb and Northrop 1916, 1917). There has been some recent excitement over "longevity mutants" (reviewed by Kenyon 2005), which also appear to exemplify the stretching pattern in that they suffer reduced metabolic rates (Van Voorhies and Ward 1999; Van Voorhies 2002) or diminished reproductive and competitive capacities (Van Voorhies 1992; Marden et al. 2003; Jenkins et al. 2004). The insulin-like signaling pathway, in particular, appears to modulate the allocation of nutrients between maintenance of the adult soma and reproduction (Kenyon et al. 1993; Chen et al. 1996; Bohni et al. 1999; Clancy et al. 2001; Tatar et al. 2001; Bartke 2005). A lot of attention has been focused on this pathway because it is conserved across distantly related species, from nematodes to fruit flies to mammals (Fontana et al. 2010).

THE SCIENTIFIC STUDY OF THE FIRST TWO PHASES OF LIFE: DEMOGRAPHY AND EVOLUTIONARY BIOLOGY

From the standpoint of demographically focused evolutionary biology (e.g., Charlesworth 1980, 1994; Rose 1991; Roff 1992; Stearns 1992), the first two phases of life have quite distinct evolutionary properties. Development takes place during a period of intense natural selection, which is how evolutionary biologists characteristically explain the relative perfection of developmental processes. Alleles that have deleterious effects on fitness-related characters early in life are not favored by natural selection unless those same alleles have beneficial effects on other parts of the life cycle. This can occur in cases of antagonistic pleiotropy between components of the life history during development (Rose 1982), one scenario being genetic trade-offs between growth rate and viability. The ways in which such trade-offs arise, and their evolutionary consequences, will often involve specific aspects of the ecology and developmental biology that affect selection in each species. As such, they are not amenable to general theoretical characterization or analysis, although some attempts were made in the older "optimal life-history" literature (reviewed in Charlesworth 1980).

The situation is quite different with respect to aging during adulthood. When demographic information is incorporated into evolutionary genetic theory, as in Charlesworth's (1980) classic monograph, then it is possible to derive predictions concerning the evolution of age-specific life-history characters. At the core of this type of theory are Hamilton's (1966) twin forces of natural selection, with one force acting on age-specific survival and the other force acting on age-specific fecundity. Both of these forces suggest that adulthood should be marked by persistent declines in age-specific survival probabilities and average fecundity. The use of this theory has led to an outbreak of interesting research on the evolution of aging and kindred topics in life history (Rose 1991; Roff 1992; Stearns 1992; Rose et al. 2007). But that is not our main focus here.

THE DEMOGRAPHY AND EVOLUTIONARY BIOLOGY OF LATE LIFE

Our concern in this volume is to elucidate the fundamental demographic and evolutionary properties of late life, concentrating on data collected from well-defined laboratory experiments. There are ongoing studies of late life using data from human cohorts. However, we do not consider such data of sufficient quality to be useful in deciding basic scientific questions. Most of these human cohort studies are demographic in nature and only provide a post hoc analysis of death and the incidence of diseases. We will discuss these problematic data in Chapter 11, as well as applying our general conclusions to the human case.

For the time being, let us just mention that the observation that mortality rates follow distinctly different trajectories in late life in humans, and do not continue to increase exponentially at very late ages, is not a new finding for demographers and actuaries studying human data (Greenwood and Irwin 1939; Comfort 1964; Finch 1990; Gavrilov and Gavrilova 1991). Much of the work being done on late ages in humans is focused on constructing and analyzing life tables and describing trends in age-specific mortality patterns (e.g., Kannisto 1994; Christensen and Vaupel 1996). What these data have revealed is that there is a slowing in the acceleration of mortality rates at around age 80, followed by a plateau after age 105 (Vaupel et al. 1998; Young et al. 2009). This slowing in mortality at late ages has been more pronounced since 1950 in developed countries (Kannisto et al. 1994; Vaupel 1997) and contributes to the increase observed in the maximum lifespan (Wilmoth et al. 2000). Other work is comparative in nature, using autopsies of the oldest old (e.g., Bernstein et al. 2004), or comparing cohorts of young or middle-aged humans with older individuals. Perhaps the most revealing studies on late life in humans are longitudinal studies that consider social, behavioral, biological, and environmental factors across the lifespan of many individuals. One such study is the Chinese Longitudinal Healthy Longevity Survey, which has produced over 60 peer-reviewed articles from the data that have already been collected (see Yi et al. 2008).

But for the present purpose, data collected from other animal species are of much greater interest. There has been a recent increase in studies of late life using a variety of laboratory animal species, especially after the definitive discovery of the leveling of age-specific dipteran mortality rates at late ages in the Carey and Curtsinger laboratories in 1992 (Carey et al. 1992; Curtsinger et al. 1992). Since the publication of their reports of late-life mortality-rate plateaus in large cohorts of medflies and fruit flies, several laboratories have investigated the late-life demographic properties of a variety of organisms (reviewed by Charlesworth and Partridge 1997; Vaupel et al. 1998; Carey 2003). To our knowledge, late-life investigations have revealed similar late-life plateauing of mortality rates in all organisms that have sufficiently large cohort numbers surviving into the aging period. Among the species that have been studied are the medfly *Ceratitis capitata* (Carey et al. 1992; Carey 2003), the commonly studied laboratory fruit fly *Drosophila melanogaster* (Curtsinger et al. 1992; Fukui et al. 1993; Clark and Guadalupe 1995; Fukui et al. 1996; Promislow et al. 1996; Vaupel et al. 1998; Drapeau et al. 2000; Rose et al. 2002; Miyo and Charlesworth 2004), the Mexican fruit fly *Anastrephan ludens* (Vaupel et al. 1998; Carey et al. 2005), a parasitoid wasp *Diachasmimorpha longiacaudtis* (Vaupel et al. 1998), the nematode *Caenorhabditis elegans* (Brooks et al. 1994; Vaupel et al. 1994, 1998; Johnson et al. 2001), baker's yeast *Saccharomyces cerevisiea* (Vaupel et al. 1998), and the beetle *Callosobruchus maculates* (Tatar et al. 1993).

The key to studying late life is cohort size. Investigators must employ enough individuals in a study so that a significant number of individuals from a particular cohort are alive at late ages. One reason that late-life mortality-rate plateaus had not been definitively observed before 1992 in the many laboratory studies of aging may have been that almost all of these studies used cohort sizes of only 100–200 individuals or less per population. Thus, the chance of enough individuals surviving to late enough ages for a plateau in age-specific mortality rates to be clearly defined was quite low.

The studies cited above collectively have demonstrated that late-life plateaus in mortality rates are thus far widely observed among experimental cohorts in which enough individuals survive well into the aging phase and the environment is maintained with stable conditions. Interestingly, late-life mortality-rate plateaus are also observable under a variety of environmental and genetic conditions. They have been observed in cohorts of inbred and outbred individuals, in genetic mutants having extended lifespans, and in cohorts kept at varying densities (see above citations). These results collectively suggest that mortality-rate plateaus are a robust finding, and that late life is a phase of life very different from both development and aging, but a phase of life requiring careful study especially in large cohorts maintained under good conditions.

However, some of these data are probably subject to artifacts of inbreeding and genotype-by-environment interaction, which complicate the interpretation of any study of demography or life-history evolution. For instance, it has long been known that subtle environmental effects, such as past density history, can affect age-specific mortality (Pearl et al. 1927). Genetic correlations will also change in different environments (Service and Rose 1985). Inbreeding has been shown to affect life-history traits in a variety of ways. Inbreeding may change genetic correlations between life-history traits (Rose 1984a), accelerate senescence (Mueller 1987), and alter population dynamics if it causes reductions in female fecundity (Prasad et al. 2003). Therefore, we will be concentrating primarily on our own experimental research, which has used *Drosophila* populations that are well adapted to our laboratory environment while being kept relatively free of inbreeding (Rose et al. 2004).

A further limitation on the present discussion is that we are not going to consider the physiological or mechanistic foundations of late life in great detail. We feel that this is a tremendously important subject, but too little is known about it at present to offer more than the preliminary findings that we will describe in Chapter 10. But we will return to this topic in future publications once we have analyzed sufficient physiological data from late life.

Thus, the present volume constitutes an intensive analysis of the demographic and evolutionary foundations of late life in theory and in well-defined laboratory cohorts, primarily from our own *Drosophila* laboratories. In this sense, then, we are covering specifically the *evolutionary biology of late life*. Given the novelty of this research area and its surprising features, we are confident that the limited scientific terrain that we survey nonetheless will be of great interest for students and scholars of evolutionary biology, demography, ecology, and gerontology.

Discovery of Late Life

Human demographers have long noticed and documented a reduction in the acceleration of human mortality rates with age. But this pattern in human data was not considered of general scientific importance until the virtual cessation of aging was documented quantitatively in two insect species in 1992. Since then, post-aging late life has been documented in a variety of experiments. These plateaus are not artifacts arising from inbreeding, density, etc. In the early 2000's, it was also discovered that late-life fecundity plateaus as well.

INTIMATIONS OF HUMAN LATE LIFE

Demographers have traditionally characterized adult age-specific mortality rates in terms of the Gompertz equation, first intuited in the nineteenth century (Gompertz 1825). This equation is usually presented in the following form,

$$\mu(x) = A \exp(\alpha x), \tag{2-1}$$

where A is the age-independent mortality parameter and α is the age-dependent parameter. The parameter α is often interpreted as reflecting the rate of aging. It is interesting that in populations of *Drosophila* selected for postponed aging, the magnitude of α has declined relative to controls, as has the magnitude of A (Nusbaum et al. 1996). On the other hand, A reflects background sources of mortality, which don't change fundamentally with age. Environmental factors, like caloric restriction or exposure to urea, increase longevity in *Drosophila* by decreasing the value of A (Nusbaum et al. 1996; Joshi et al. 1996).

There are a number of variants of this model, like the Gompertz-Makeham model,

$$\mu(x) = R + A \exp(\alpha x). \qquad (2\text{-}2)$$

Equation 2-2 was later revised by Makeham to include both linear and exponential components (see Gavrilov and Gavrilova 1991),

$$\mu(x) = R + Sx + A \exp(\alpha x). \qquad (2\text{-}3)$$

None of these variants of the Gompertz equation have profound biological motivations. However, all of these models have in common the assumption that age-specific mortality rates increase with positive acceleration, assuming positive parameter values, as is conventional. For example, the mortality rate of humans between ages 50 and 60 years will be underestimated if it is taken as a simple continuation of the mortality rate between 40 and 50 years of age.

Gompertzian models, a term that we use to refer to the entire class of models under one rubric, often fit mortality rate data extremely well (e.g., numerous plots in Finch 1990). However, there was no profound scientific justification for Benjamin Gompertz's (1825) original proposal of models of this kind. It was merely the simplest of an entire class of models that might be fit to actual experiment data.

The post hoc nature of this model for human aging was revealed by the indifference shown by scientists to the discovery of substantial slowing in human age-specific mortality rates at later ages. Patterns of this kind were casually noticed in the nineteenth century, but it wasn't until 1939 that Greenwood and Irwin analyzed abundant human European data to show that human mortality rates essentially plateau late in life, as we discuss in detail in Chapter 11. Because there was no particular theory underlying the Gompertz equation and its Gompertzian congeners, this anomaly in human demographic data did not provoke any direct scientific examination of the departure of late life from the Gompertzian pattern of the aging phase of human demography. Some scientists were willing to attribute the apparent slowing, if not cessation, of demographic aging to recondite and incidental factors, such as improved care in nursing homes, among other changes of behavior and environment in patients who are extremely old (e.g., Olshansky et al. 1993; Maynard Smith et al. 1999, p. 269). Although a better standard of living, public health measures, and medical developments can all contribute to the slowing of mortality rates in the oldest old humans, these factors should also slow mortality rates at all other ages.

The fundamental problems with human data concerning age-specific mortality are several. Humans as individuals are aware of their aging, both immediately and longitudinally, which may cause them to change their behavior with time. Humans as social animals are subject to age-dependent social interactions. Young humans are treated one way, mature reproductive adults another, and postreproductive adults yet another. This is unquestionable. Humans are very long-lived, which makes full longitudinal studies of our lives hard to arrange and carry out. Humans have lived through remarkably different epochs over the last few centuries, as the pace of global change has been striking, affecting our nutrition, exposure to disease, and medical treatment. In general, a worse experimental system for distinguishing among demographic epochs within a life cycle is hard to imagine, despite the vast amounts of data concerning human age-specific mortality patterns.

THE REVOLUTION OF 1992: DEFINITIVE EVIDENCE FOR A THIRD PHASE OF LIFE

Scientific studies of age-specific survival rates in nonhuman species have suffered in other respects. The problems with inferring underlying endogenous patterns of mortality from wild populations are obvious. Most wild animals move around, making the accurate estimation of their mortality extremely difficult, because missing animals might have left the study area or be unrecoverable or undetectable for other reasons, without having died. Wild plants do not move around, but like animals, they are subject to environmental vagaries, from weather to grazing to disease to wildfires.

The problems with studies of laboratory populations are somewhat more recondite but equally substantial. Most studies of mortality in laboratory cohorts use small numbers. While human mortality data from Europe may involve millions of individuals and studies of wild populations of other species may involve thousands of individuals, laboratory studies of mortality in animals or plants are more likely to involve only dozens to hundreds of individuals. Furthermore, many standard laboratory stocks are either recently introduced to the laboratory or highly inbred. Genotype-by-environment interaction makes the properties of newly introduced organisms highly unpredictable (vid. Matos et al. 2004; Teotónio et al. 2004; Rose et al. 2005), and such data are not useful for the interpretation of the survival rates of organisms in their normal habitats. Highly inbred organisms from species that do not usually self-fertilize are subject to inbreeding depression, which can produce highly anomalous life-history data (Rose 1984a, 1991).

The combined effects of these problems afflicting human, wild, and laboratory studies of age-specific mortality are fairly devastating. The amount of appropriate data on the *lifelong* pattern of age-specific mortality is very limited. Furthermore, the data degrade with age in all finite cohorts, simply because there are fewer individuals alive at later ages, making the sampling variation of age-specific mortality increase rapidly with age at very late ages.

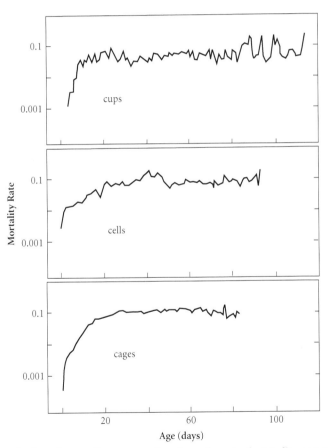

Figure 2-1. The log of mortality rate as a function of age in the Mediterranean fruit fly, *Ceratitis capitata.*
SOURCE: Carey et al. 1992.

Given all these difficulties, it is not surprising that it was not until 1992 that the first scientifically definitive data on late life were published by Carey et al. and Curtsinger et al. The scientific importance of these papers is hard to over state. Though many quibbled over their significance, including some of the present authors (e.g., Graves and Mueller 1993; Nusbaum et al. 1993), in retrospect, they can be seen as two of the most revolutionary scientific papers published in the twentieth century.

The keys to their value were scale, standardization, and replication. Millions of flies were used to estimate age-specific mortality rates among adults. Great care was taken to handle these organisms in standardized ways, as opposed to the haphazard nature of the data from human or wild population studies of similar size. Handling methods were varied among experiments, and some of the experiments were replicated internally as well, with multiple cohorts handled in parallel at the same time.

Carey's laboratory collected data for three different kinds of cohorts of medflies (Carey et al. 1992). In two of these experiments, cohorts of more than 20,000 individuals were housed individually in either cups or tissue cells, the flies having more room in the former type of housing compared to the latter, but still with relatively little opportunity for much activity in either case (Figure 2-1). Thus, many of the normal environmental hazards associated with aging, like mating, egg laying, activity, and density effects, were limited. The third cohort study by Carey et al. comprised over 1.2 million medflies that were housed in cages containing approximately 7,200 flies each. This cohort, unlike the other two, experienced mating, egg laying, activity, and decreases in density with age. Despite these differences, all three cohorts demonstrated a similar plateauing of age-specific mortality rates at later ages (Figure 2-1).

Curtsinger's group observed a similar leveling of age-specific mortality rates at late ages in a single genotype of male *Drosophila* (Figure 2-2, data from Curtsinger et al. 1992). Their experiment employed an inbred cohort of 5,751 male fruit flies to better understand the details of mortality rates at very late ages. They found that a Gompertz-type mortality model fit the data quite well until 30 days, after which mortality rates were better fit to a constant mortality rate (Figure 2-2).

Mortality plateaus have been documented in several other species, including houseflies (Rockstein and Lieberman 1959), bruchiid beetles (Tatar et al. 1993; Tatar and Carey 1994a,b, 1995), two different species of seed-feeding beetles (Fox et al. 2006), and butterflies (Gotthard et al. 2000).

To summarize the forgoing studies, these data show that the process of exponentially increasing age-specific death rates can come to an end under well-defined laboratory conditions, given cohort sizes large enough to allow accurate estimation of age-specific mortality rates late into adult life. After the point of detectable mor-

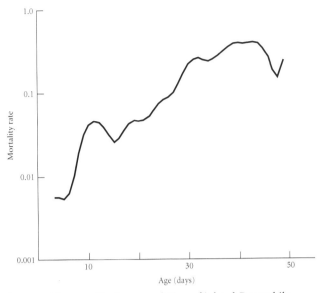

Figure 2-2. Age-specific mortality in a population of inbred *Drosophila*.
SOURCE: After Curtsinger et al. 1992.

tality-rate deceleration, mortality rates seem to roughly stabilize, with some cohorts even showing *declining* age-specific mortality rates. However, in principle, a stable *average* mortality rate is expected to lead to some *sample paths* with declining mortality rates, so individual instances of such declining mortality rates are not necessarily important. In other words, in well-defined laboratory cohorts, the human late-life pattern of decelerating aging was found again.

The initial publications revealing decelerating aging in laboratory animal species were met with some skepticism and scrutiny. Many critics offered explanations for the observed late-life plateaus in mortality, explanations such as the reliability theory of aging (Gavrilov and Gavrilova 1993), heterogeneity effects (Kowald and Kirkwood 1993), population density effects (Nusbaum et al. 1993), and age-related changes in activity and sample size (Olshansky et al. 1993). Although Carey et al.'s experiments certainly addressed the problem of small sample size and Curtsinger et al.'s study addressed the effect of genetic heterogeneity on the leveling of late-life mortality rates, it was clear that further experiments were necessary.

Curtsinger's laboratory addressed a number of these concerns in the years following the 1992 publications. The first of these studies examined mortality-rate patterns at late ages in four inbred lines of *Drosophila*, using both males and females (Fukui et al. 1993), and thus addressed the concern that genetic heterogeneity resulted in the slowing of mortality rates at late ages. Approximately

1,000 individuals were housed in each of 18 population cages, and mortality was monitored daily until all flies had died. Flies were not recombined between cages as death occurred, so the effects of decreasing density with age were not controlled. However, this experiment further demonstrated that the leveling of mortality rates at late ages was not due to genetic heterogeneity and that a two-stage Gompertz model with a second-stage plateau best fit age-specific mortality data from large cohorts.

The next experiments in the Curtsinger laboratory studied the effects of density on age-specific mortality rates, also using four inbred lines of *Drosophila* (Khazaeli et al. 1995a). Many critics of the original studies thought it possible that the slowing of mortality rates was at least partially due to declining adult density with age (e.g., Graves and Mueller 1993, 1995; but see Curtsinger 1995a, 1995b for caveats). To address this concern, Khazaeli et al. (1995a) started three fruit fly cohorts with initial densities that varied 10-fold and then followed the mortality of approximately 70,000 flies of both sexes in total. Mortality leveled off at late ages, just as it had before, regardless of cohort density. Vaupel and Carey (1993) performed a similar test using medflies and obtained the same qualitative results.

Varying initial cohort density did not rule out the possibility that mortality rates were slowing at later ages because of a declining age-specific cohort density with age. Addressing this concern would require that the individuals within a cohort be kept at constant densities at all ages. Khazaeli et al. (1996) did just that. They observed age-specific mortality rates for two adult densities, constant at all ages, in four inbred lines of *Drosophila*. Mortality rates decelerated and plateaued at late ages even when density was held constant at all ages. Note, however, that this study did not rule out the effects of density on mortality rates or lifespan. It just showed that density effects clearly were not the cause of the mortality-rate stabilization that characterizes late life in large cohorts.

Although studies using inbred lines of *Drosophila* and outbred cohorts of medflies suggested that genetic heterogeneity was not the cause of late-life mortality-rate plateaus, environmental heterogeneity was still a possible contributing factor. Khazaeli et al. (1995b) addressed this possibility by applying an environmental stress to *Drosophila* cohorts at early ages that did not incapacitate the flies, a study referred to here as a *stress experiment*. The response of the cohort to this stress was an initial spike in mortality rates, followed by a decrease in mortality rates for the experimental cohorts compared to controls. The results of this stress experiment suggested that there is heterogeneity in mortality rates even in genetically homogeneous cohorts, and that this heterogeneity may be environmentally based. However, further analysis of the data from this experiment by the authors required them to retract these conclusions (Curtsinger and Khazaeli 1997). The theoretical and experimental work surrounding the effects of such heterogeneity effects on late-life mortality rates will be addressed in full in later chapters of this book.

One of the last experiments performed in the Curtsinger laboratory in response to the initial skepticism surrounding late-life mortality-rate plateaus again addressed the issue of inbred lines (Fukui et al. 1996). Curtsinger et al.'s original experiment with *Drosophila* used inbred lines, and only male mortality rates were observed. Although Carey et al.'s experiment with medflies used out-bred cohorts of flies, there was some concern with the design of the fruit fly experiment. Consequently, Fukui et al. (1996) studied mortality rates in both inbred and outbred cohorts of *Drosophila* for both males and females. They found that late-life mortality rate deceleration was not unique to inbred cohorts of fruit flies, or to males kept separately from females, when a large number of individuals are used.

Collectively, these experiments eliminated a number of possible artifactual reasons for the observation of late-life plateaus in mortality rates. They demonstrated that the leveling in mortality rates observed late in adult life was not a result of genetic heterogeneity, that this phenomenon was not specific to inbred lines, and that neither initial cohort density nor age-specific density decline within a cohort affected the existence of late-life mortality-rate plateaus. Plateaus in mortality rates were well established as a robust finding needing much more theoretical and experimental attention by the later 1990s.

THE REVOLUTION CONTINUED WITH FECUNDITY

Rauser et al. (2003) intuited that late-life fecundity would also show a cessation of aging similar to mortality and tested whether this was in fact true. In evolutionary biology, *aging* is defined as a sustained endogenous decline in age-specific fitness components (Rose 1991). These fitness components include characters other than age-specific mortality. Most importantly, they include such characters as age-specific female fecundity, age-specific male mating success, and the like. Thus, if aging generally ceases at late ages, then it should cease for these characters as well, leaving aside organisms that undergo such distinct reproductive terminations as human menopause.

As predicted, female fecundity in *Drosophila* shows a gradual decline with adult age, with some degree of deceleration in the rate of decline toward the end of adult life. To illustrate this deceleration, we have displayed the age-specific fecundity from replicated measurements of five outbred *D. melanogaster* populations called CO_1 to CO_5 (Figure 2-3). These populations are cultured with moderate selection for late-life fitness (reproduction is at 4 weeks of life, about 16–18 days from the onset of adulthood). The results shown in Figure 2-3 are taken from six different experiments: a comparison of the five CO populations with a set of five populations called the ACO (Rauser et al. 2006b), a test of lifelong heterogeneity on female fecundity plateaus (Rauser et al. 2005a), a test of the effects of male age on fecundity plateaus (Rauser et al. 2005b), a test of nutrition level on fecundity plateaus (Rauser et al. 2005b), a comparison of the CO populations and a set of five reverse-selected populations, the NRCO's (Rauser et al. 2006b), and a comparison of fecundity and mortality plateaus (Mueller et al. 2007). This is the largest set of lifelong age-specific fecundity data collected under good laboratory conditions known to us. While there is significant variation in the specific patterns of individual experiments, as a group these fecundity trajectories show a clear deceleration in the rate of decline in age-specific fecundity. Furthermore, the average plateau heights that we have found are at levels statistically greater than zero. That is, fecundity levels do not merely plateau by achieving a value of zero.

It is important to note here that when we are discussing plateaus in fecundity, we are referring to the fecundity patterns of a population, not the fecundity patterns of individual females. Note also that there is a significant complication facing fecundity due to the effects of the process of dying. We analyze the effects of this *death spiral* in Chapters 3 and 9.

Despite the consistency of fecundity-decline deceleration in data like these, we were concerned about several possible artifacts, specifically whether male age or high nutrition causes the cessation of reproductive aging in females (Rauser et al. 2005b). That is to say, upon our initial discovery of late-life fecundity plateaus, we were as concerned as the Curtsinger laboratory investigators were about possible artifacts that might have generated a misleading appearance of a distinct late-life phase following aging. As our tests for these possible artifacts have not been as widely communicated as the earlier work of the Curtsinger laboratory vindicating late-life mortality rate plateaus, we review this *Drosophila* fecundity research in some detail here.

ARE FECUNDITY PLATEAUS CAUSED BY INADEQUACY OF OLDER MALES?

The first of these possible artifacts is diminished male sexual function. The idea is that a slower rate of egg depletion, and thus a stabilization of later-life fecundity, might have arisen from reduced availability of sperm among our experimental *Drosophila* cohorts. We tested this artifact hypothesis by supplying females with young males before their fecundity declined to plateau levels. In our first late-life fecundity study that suggested the existence of

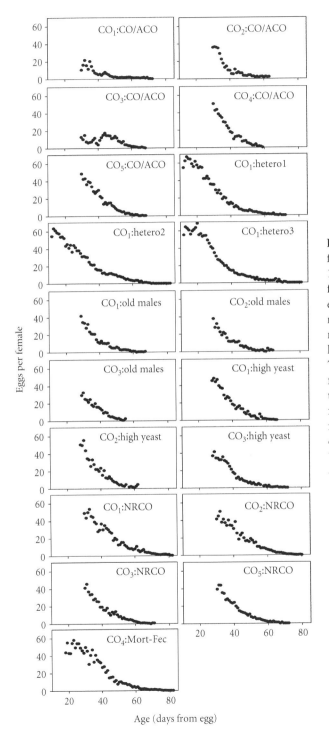

Figure 2-3. Average female fecundity in 19 CO populations from six different experiments, where the numerical subscripts refer to individual replicate CO populations. The additional coding for the graphs refers to the type of experiment from which CO cohort fecundity data were obtained, as follows: "CO/ACO" refers to an experiment comparing cohorts from CO and ACO populations; "hetero" refers to an experiment studying lifelong heterogeneity; "old males" refers to an experiment testing the environmental effect of male mate on female fecundity; "yeast" refers to an experiment testing the environmental effect of yeast level on female fecundity; "NRCO" refers to a reverse-selection experiment.

late-life fecundity plateaus, male and female cohorts were handled in parallel, without replacement, throughout life (Rauser et al. 2003). This design resulted in a supply of older mates for the older females, raising the prospect that older males may have limited female fecundity at later ages, generating an artifactual plateau in their fecundity. (We should be clear that we do not believe this hypothesis; we merely conducted a test of it.) Certainly one simple, if farfetched, explanation of our experimental results might be that late-life female fecundity plateaus may have arisen from diminished male sexual function associated with male aging, which in turn could reduce female reproduction, causing both a decline in female fecundity before the plateau and the plateau itself.

Various components of male sexual function in *Drosophila* have been shown to decline with age. Among these functions are overall mating success or ability (Aigaki and Ohba 1984; Kosuda 1985; Service 1993; Hughes 1995), which may be related to decreases in the production of both sperm and accessory gland proteins that function to elevate egg laying and increase the female death rate, among other things (reviewed by Wolfner 1997). Furthermore, Prowse and Partridge (1996) found that males always exposed to virgin females were sterile when at least 80% of their cohort was still alive. Thus, it is easy to imagine that the decline in fecundity with female age, and even the plateau itself, may be a result of an age-related decline in the sexual physiology of males.

This idea was tested by supplying females with young males before their fecundity declined to plateau levels. We reasoned that if older males artifactually created late-life fecundity plateaus, then supplying younger mates to older females should either delay the onset of the fecundity plateau until the new mates become old or obliterate the plateau altogether, because few females would survive to late enough ages for the fecundity plateau to be observable. Therefore, a supply of young males to midlife females should have caused the fecundity plateau at late ages to disappear, but only if male reproductive inadequacy established the timing and existence of fecundity plateaus. On the other hand, if male age did not cause fecundity plateaus, then we still should have observed a plateau in the fecundity of older females given young males. Note, however, that this plateau could have occurred earlier or later because of the physiological effects of supplying older females with younger mates.

To assess the effects of male age, and subsequently nutrition, on the late-life fecundity dynamics of *Drosophila,* we developed a simple model of age-specific fecundity. This model was a three-parameter, two-stage linear model, with a second-stage slope of zero, analogous to the two-stage models previously used by us to fit both adult mortality data (Drapeau et al. 2000) and fecundity data

(Rauser et al. 2003). Under such a two-stage model, the fecundity of a female aged t-days can be given as

$$\begin{cases} \varphi_1 + \varphi_2 t \; if \leq \varphi_3 \\ \varphi_1 + \varphi_2 \varphi_3 \; if > \varphi_3 \end{cases}, \qquad (2\text{-}4)$$

where φ_1 is the y-intercept, φ_2 is the slope, and φ_3 is the fecundity breakday. The term *breakday* refers to a hypothetical transition to a stable late-life condition. Note that this model does not force the data to conform to a two-stage pattern; all the data can conform to a one-stage pattern of sustained decline, given by the linear model fit to the first stage.

We also determined the height of the late-life fecundity plateau for both treatments (young and old males), along with 95% confidence intervals for these plateau heights, using the parameter estimates obtained from the two-stage linear model (Table 2-1, Rauser et al. 2005b). For both the young- and old-male treatments, the height of the plateau, or the number of eggs per female per day after the breakday, was significantly greater than zero (Rauser et al. 2005b). These results demonstrated that aging males were not artifactually causing the *occurrence* of the fecundity plateaus that we observed at very late ages in our previous experiment (Rauser et al. 2003), although the supply of younger males did affect the overall timing and shape of the plateaus.

Our test indicated that the addition of young males resulted in a more rapid onset of the fecundity plateau, as well as a plateau with an increased height (Rauser et al. 2005b). However, the earlier onset of the fecundity plateau in the young-male treatment may explain the increased height of the plateau. That is, young males caused fecundity to stop declining at an earlier age than older males, resulting in a greater number of eggs per female per day at later ages. The important conclusion from this study, however, is that female fecundity plateaued at late ages regardless of the age of their mates.

ARE FECUNDITY PLATEAUS CAUSED BY HIGH NUTRITION?

The second artifact hypothesis that we tested is that fecundity plateaus arose in our experiments from a change in the calories available for reproduction. That is, it is possible that the late-life fecundity plateaus observed by Rauser et al. (e.g., 2003) simply reflected an incidental side effect of a shift in resource allocation. The potential for such shifts in resource allocation involving fecundity is well established in the *Drosophila* populations that we used to study lifelong

Table 2-1. Parameter estimates from the two-stage linear model that was fit to mid- and late-life fecundity data from each type of CO population, those having either males of the same age as the females (old males) or young males, added when the females were age 40 days (from egg). The model was fit by nonlinear least squares regression. The height of the fecundity plateau was computed from Equation 2-4, and the estimated height was significantly different from zero ($p < 0.05$ for each population).

Population	First-stage y-int (φ_1)	First-stage slope (φ_2)	Breakday (φ_3)	Plateau height $\pm$ 95% c.i. (eggs/female/day)
Old Males				3.24 ± 0.64
CO_1	92.16	−1.94	45.23	
CO_2	77.23	−1.60	46.17	
CO_3	70.41	−1.42	48.09	
Young Males				6.89 ± 3.33
CO_1	88.61	−1.86	43.23	
CO_2	101.21	−2.34	39.67	
CO_3	76.79	−1.61	46.12	

NOTE: Parameter estimates for φ_1, φ_2, and φ_3 were all significantly different from zero; $p < 0.0001$ for each of the three populations under both treatments.

fecundity, with decreased food strongly associated with reduced fecundity and increased longevity (Chippindale et al. 1993, 1997).

In particular, it is conceivable, although somewhat implausible, that the fecundity plateau phenomenon may have been generated by artifacts arising from environmental factors that modulate female reproduction that arose specifically in the study of Rauser et al. (2003) but are not general. In our first experiments that established the existence of late-life plateaus in fecundity (Rauser et al. 2003), females were provided high levels of yeast, following the protocols of Chippindale et al. (1993), throughout the duration of the assays. Female fecundity may have plateaued later in life only because of such sustained high nutrition at all earlier ages. Therefore, we were concerned that females maintained at a lower level of nutrition might not exhibit a late-life plateau in fecundity levels.

Furthermore, fecundity is generally known to be highly responsive to nutrition in *Drosophila*. Specifically, egg laying, utilization of sperm, mating frequency, and vitellogenesis all increase with increasing nutrition

(Chippindale et al. 1993; Chapman et al. 1994; Chapman and Partridge 1996; Good and Tatar 2001). However, this increase in overall female reproduction does not come without a cost (cf. Partridge 1987; Reznick et al. 2000) and may even be coupled with a decrease in survival (Fowler and Partridge 1989; Chapman et al. 1993, 1994, 1995; Chippindale et al. 1993; Chapman and Partridge 1996).

Such plateau-disappearance and resource-diversion hypotheses were tested by comparing the effects of high and low nutrition levels on female fecundity at later ages. Specifically, we supplied cohorts derived from each experimental population with either high nutrition or low nutrition and measured fecundity throughout their adult lives. If high nutrition artifactually allows fecundity to plateau at late ages, but plateaus do not otherwise occur, then supplying flies with low nutrition should eliminate the plateau altogether. On the other hand, if nutrition does not affect the *existence* of fecundity plateaus, then we should still observe a plateau in the fecundity of females given low nutrition. Note, however, that we were not testing whether or not nutrition affects fecundity at all. That point is well established in the experimental literature. Instead, we were testing whether late-life fecundity plateaus continue to arise as nutrition is experimentally varied.

As with the previously described artifact test, we also determined the height of the late-life fecundity plateau for both treatments (high and low nutrition), along with the 95% confidence intervals for those heights, using the parameter estimates obtained from the two-stage linear model described above (Table 2-2, Rauser et al. 2005b). We found that the height of the fecundity plateau was significantly greater than zero in the presence of both high and low nutrition. As in the young males experiment, fecundity plateaus arose regardless of nutrition level at some number of eggs greater than zero (Rauser et al. 2005b).

Although the existence of fecundity plateaus was not affected by varying nutrition levels, fecundity was lower under low nutrition compared to high nutrition at all ages, including those ages after the onset of the plateau. Furthermore, fecundity declined at a much slower rate under low nutrition compared to high nutrition (see φ_2, Table 2-2). These results are consistent with dietary restriction and other nutrition studies, which have demonstrated that low nutrition levels result in decreased daily and lifetime fecundity in *Drosophila* (David et al. 1971; Trevitt et al. 1988; Chippindale et al. 1993; Chapman and Partridge 1996; Good and Tatar 2001). However, the result that the plateau height was significantly different from zero regardless of nutrition level allowed us to reject the artifact hypothesis that late-life fecundity plateaus were caused merely by high nutrition in the experiment of Rauser et al. (2003).

Table 2-2. Parameter estimates from the two-stage linear model that was fit to mid- and late-life fecundity data from each experimental cohort having either high nutrition (5.0 mg/vial) or low nutrition (0.2 mg/vial) throughout each assay. The model was fit by nonlinear least squares regression. The height of the fecundity plateau was computed from Equation 2-4, and the estimated height was significantly different from zero ($p < 0.05$ for each population).

Population	First-stage y-int (φ_1)	First-stage slope (φ_2)	Breakday (φ_3)	Plateau height ± 95% c.i. (eggs/female/day)
High Nutrition				3.43 ± 2.38
CO_1	106.24	−2.11	48.39	
CO_2	113.00	−2.30	46.90	
CO_3	91.35	−1.80	48.57	
Low Nutrition				0.82 ± 0.56
CO_1	46.17	−0.64	70.98	
CO_2	39.63	−0.55	70.42	
CO_3	29.25	−0.38	72.65	

NOTE: Parameter estimates for φ_1, φ_2, and φ_3 were all significantly different than zero; $p < 0.0001$ for each of the three populations under both treatments.

CONCLUSION: A THIRD PHASE OF LIFE HAS BEEN EXPERIMENTALLY ESTABLISHED

From the standpoint of the demography of aging, in particular its eventual cessation, our fecundity results are comparable to the finding that mortality rates plateau. Both findings have now been tested for possible artifacts, and somewhat replicated, especially in *Drosophila*. Together, they suggest that aging is demographically a transition between two periods of relatively stable mortality and fecundity levels, which may render it more amenable to eventual control than a process that is an endless and exponential rise in age-specific mortality, debility, and sterility.

Is late life a period in which all age-specific life-history characters stabilize? We are not aware of comparable data for male age-specific fitness components, the third major type of adult life-history character. However, it is now reasonable to propose the bare hypothesis that *aging generally ceases in cohorts of organisms that live long enough as adult somata under benign and stable condi-*

tions. But without doubt, after a period of persistently increasing age-specific mortality and decreasing age-specific reproductive output, a third phase of life *can* occur, a phase that is clearly unlike aging demographically. It is this stunning finding that animates our work on late life, whether or not this cessation of aging is indeed universal. In particular, we have sought not merely to document this fact. We have also tried to explain it in terms of basic biological theory, this explanation being the chief topic of the present book.

Late Life Is Predicted by Hamiltonian Evolutionary Theory

Hamilton's forces of natural selection acting on age-specific survival and fecundity imply strong selection during development and weakening selection during the first part of adulthood. After the last ages of survival and reproduction in a population's evolutionary history, Hamiltonian theory predicts endless plateaus in the forces of natural selection. Explicit models of life-history evolution show that these plateaus allow the evolution of late-life plateaus in both age-specific survival and fecundity.

THE FORCE OF NATURAL SELECTION ACTING ON MORTALITY: FROM AGING TO LATE LIFE

Haldane (1941) and Medawar (1946, 1952) were the first to describe aging as a consequence of the weakening force of natural selection with age. However, it was Hamilton (1966) who actually derived a quantitative formula for this force for the first time. According to Hamilton, the force of natural selection acting on mortality is given by $s(x)/T$, where x is chronological age and T is a measure of generation length. The function s at age x is given by

$$s(x) = \sum_{y=x+1} e^{-ry} l(y)m(y), \tag{3-1}$$

where r is the Malthusian parameter, or the growth rate of the population, associated with the specified $l(y)$ survivorship and $m(y)$ fecundity functions. The variable y is used to sum up the net expected reproduction over all ages after age x. Ultimately, the $s(x)$ function represents the immediate fitness impact of an individual's future reproduction. Note that, before the first age of reproduction (b), s is always equal to 1 (one): once reproduction has ended, s is equal to zero; and during the reproductive period, $s(x)$ progressively falls. Figure 3-1 shows an example of an $s(x)$ function that depicts how the force of natural selection acting on mortality declines with adult age, throughout the reproductive phase, and converges on zero at very late ages.

From 1966, when Hamilton first published his analysis, until 1996, it was generally assumed by evolutionary biologists that the decline toward zero values of Hamilton's force of natural selection acting on age-specific mortality rates implied unremitting increases in age-specific mortality rates (e.g., Rose 1991). This was historically significant, because it was thus assumed that evolutionary theory provided a fairly direct warrant of the practically universal intuition among biologists that they could interpret aging as effectively a collapse in the working of a biochemical machine that had worked efficiently at an early age. In effect, this intuition led many gerontologists to the view that they could largely ignore evolutionary issues in their research and just focus on patterns of cumulative breakdown (vid. de Grey and Rae 2008).

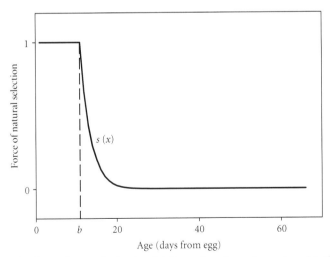

Figure 3-1. The force of natural selection ($s(x)$) as a function of age where b is the first age of reproduction.

The research that we will now present supplies an argument that the intuitive interpretations of all biologists prior to 1996 were incorrect, from the intuitions of evolutionary biologists to those of gerontologists. Consider the following possibility. What if the roughly exponential rise in age-specific mortality rates with adult age might instead be a direct reflection of the pattern of change in the force of natural selection acting on age-specific mortality, and not just its decline to lower average values? If this intuition is correct, then it implies that later plateaus in age-specific mortality could arise from the plateauing of the force of natural selection acting on mortality.

In order to test the formal cogency of this alternative interpretation of Hamilton's forces, Mueller and Rose (1996) set about simulating the evolution of age-specific mortality using standard equations for the evolution of allele frequencies in age-structured populations. The results of those simulations corroborated the alternative interpretation: in every case that they examined, regardless of the pattern of pleiotropy or mutation that they employed, late-life plateaus evolved in the simulated populations of Mueller and Rose (1996).

Put simply, implicit in Hamilton's (1966) original theory for the force of natural selection is an evolutionary theory of late life. Recall that s is equal to zero for all ages after reproduction has ceased. Therefore, it is intuitive that age-specific mortality rates should mimic the plateau in the force of natural selection, because natural selection is unable to distinguish fitness differences in survival at different ages after the cessation of reproduction in the course of a population's evolution. Survival rates do not necessarily have to reach zero as soon as reproduction ceases, because beneficial effects that are not age-dependent will continue to benefit individuals who remain alive after the force of natural selection has converged on zero. Any age-independent genetic benefits will be favored by natural selection acting at early ages and will have positive pleiotropic benefits at all later ages.

A SIMPLE ANALYTICAL EXPLANATION OF LATE-LIFE MORTALITY PLATEAUS

Before we turn to examples of simulation results for the evolution of late life, we will provide a simple mathematical sketch that may help some readers understand the qualitative features of late-life evolution. Suppose that age-specific survival in a particular environment is an exponentially decreasing function of age $(l_x = e^{-dx})$ and fecundity is an exponentially increasing function of age $(m_x = e^{fx})$. Then for a population with a stable age distribution where

$$\sum_x e^{-rx} l_x m_x = 1,$$

(3-2)

one can show that $f-r-d \leq 0$, and so the magnitude of the terms in Equation 3-2 is exponentially decreasing to zero. In fact, this result should hold for any schedule of l_x and m_x since we can always find a d such that $e^{-dx} > l_x \forall x$ and, likewise, we can find an f such that $e^{fx} > m_x \forall x$ and thus, $e^{-rx} l_x m_x < e^{x(f-r-d)} \to 0$ as $x \to \infty$. Consequently, there should always be an age at which the force of selection is so small that it has a trivial impact on allele frequency dynamics relative to forces like random genetic drift. And for all ages after that, the force of natural selection will remain negligible. This implies that natural selection should be entirely unresponsive to *differences in the age(s)* at which alleles affect either age-specific survival or reproduction during this later part of life, making such alleles adaptively equivalent. Thus, our first-order expectation is that forces of natural selection will generally plateau in such a way as to produce a "plateau" of adaptation at sufficiently late adult ages.

NUMERICAL SIMULATIONS OF THE EVOLUTION OF LATE-LIFE MORTALITY

Now we will present some basic numerical results that quantitatively show the patterns that the Hamiltonian theory of age-specific selection can produce for age-specific survival. To develop these quantitative results, we have carried out extensive numerical simulations. But before presenting the results of these simulations, it is important to address the types of inference that we believe are possible from this type of theoretical work.

All computer simulations suffer from the problem that they can only produce a finite number of specific results. Given these results, it is generally hoped that through induction we can infer some general patterns that hold beyond the specific cases examined. The advantage of simulations is often that complicated problems that are not easily amenable to standard mathematical analysis can be examined without the need to make the many simplifying assumptions that such analysis requires, assumptions that are typically unrealistic and chosen strictly for their mathematical convenience. Mathematically tractable theory will often have limited generality due to these simplifying assumptions.

One standard pattern of mortality that characterizes many well-studied organisms is a pattern of exponentially increasing mortality with age that is sometimes described by the Gompertz equation. In this chapter, we develop evolutionary outcomes in populations that are initially assumed to have a Gompertz mortality pattern. By allowing mutations with small effects to alter the basic mortality pattern, we develop numerical examples that help reveal the stability over short-term evolution of the Gompertz mortality pattern.

These simulations share many of the attributes of a modeling approach called *adaptive dynamics* (Waxman and Gavrilets 2005). We assume that a population is initially monomorphic for a life history characterized by a survival (l_x) schedule and a fertility (m_x) schedule. A mutant with an altered survival schedule ($\tilde{l}_x$) is then introduced into the population. The initial increase and ultimate fixation of this mutant are determined by comparing the fitness of the mutant ($\tilde{r}$) to the fitness of the resident (r) genotype, where r is determined from Equation 3-2 and $\tilde{r}$ is determined from a similar equation with $\tilde{l}_x$ substituted for l_x. As discussed by Waxman and Gavrilets (2005), the conditions required for the initial increase of a mutant may not be the same as the conditions required for fixation. For instance, this procedure ignores genetic complications like stable polymorphisms.

Perhaps of greater concern is the manner in which new mutants are generated. Three basic models of mutation are used in many evolutionary models: (1) a continuum of possible phenotypes that are measured as departures from the resident phenotype, with larger departures being less likely (Crow and Kimura 1964); (2) the house of cards model, in which the distribution of mutant effects is statistically independent of the resident phenotypes (Kingman 1978); (3) the regression model of mutation, which is intermediate between the house of cards model and the continuum model (Zeng and Cockerham 1993).

The house of cards model is based on the logic that most mutations are deleterious and thus produce a phenotype unlike that of the resident. While this is generally a reasonable assumption, in these adaptive dynamic models widely deleterious mutants would be eliminated quickly, making them impediments to the simulation of adaptive evolution. The successful mutants that actually improve fitness are more likely to be only small changes from the resident phenotype. Accordingly, we have used the continuum of mutation model in these simulations.

Generation of Mutants

We assumed a fixed maximum lifespan of 109 days and a 9-day development time, giving 100 discrete adult age classes and nine juvenile age classes. (This is a dipteran kind of life cycle; formally, it can be made into a human-scale life cycle by converting all age classes and genetic effects into approximately 1-year time units, with the development period somewhat lengthened.) We considered a target period or window for mutational effects of fixed duration. Here we show simulations using a period of 10 age days. These periods correspond to the pleiotropic effects of these mutations. Many critiques of our original theory (Mueller and Rose 1996) have focused on the special case of mutants that affect only a single age class and thus were lacking these pleiotropic effects (Pletcher and Curtsinger 1998; Wachter 1999).

The ages at which these effects were imposed were chosen at random. Each mutation had a window of beneficial effects and a window of deleterious effects. The first day of action was chosen independently for the beneficial window and the deleterious window. Windows that would exceed the oldest age class were truncated to have their last effect at adult age 100 days. For each age class in the window of beneficial effects, the new probability ($\tilde{P}_x$) of surviving from the current (x) to the next ($x + 1$) age class is given by

$$\tilde{P}_x = P_x + (1 - P_x)\frac{\delta}{\omega}, \tag{3-3}$$

where δ is a constant (set to 0.1 in these simulations), ω is the number of days in the window (set to 10 in these simulations), and P_x is the age-specific survival of the current resident genotype. Deleterious effects were assumed to result in a new age-specific survival value:

$$\tilde{P}_x = P_x \left(1 - \frac{\delta}{\omega}\right). \tag{3-4}$$

We initially chose 10,000 pairs of random days for the onset of the beneficial and deleterious days of each mutant. The order of the 10,000 pairs was then randomly shuffled into 100 different vectors. Each vector constituted a different ordering of these mutations. To some extent, the outcome of evolution may be affected by the order in which these mutants are introduced into the population. Thus, the results that we give here consist of 100 realizations of evolution from the same ancestral populations. We used these 100 realizations to construct 96% confidence intervals on the changes in age-specific mortality. Descriptions of additional properties of these simulations can be found in the Appendix section for this chapter. This Appendix section also reviews results from alternative models for generating mutants. These alternative models also yield plateaus according to our calculations, and thus we believe that the basic results presented in the next section are not artifacts of special assumptions made here.

Relative Roles of Drift and Selection

Our approach is to study the evolution of life histories in the vicinity of a reasonable starting life history. However, there is some interest among theoreticians in understanding the long-term behavior of models, which is sometimes difficult to determine from computer simulations, because simulated dynamics can be too slow to provide an adequate guide to asymptotic behavior. Nevertheless, in the appendix, we give some examples with short lifespans where in fact we can demonstrate that the populations have converged to a locally stable equilibrium. We do this by simply computing the fitness of all possible new mutants as evolution proceeds, such that when it is impossible to generate a mutant with greater fitness then the current resident we conclude the current resident genotype is the equilibrium phenotype.

It seems reasonable to suppose that, if equilibria can be identified for short life cycles, they should also exist for longer life cycles. However, these equilibria will certainly take longer to reach in the latter case. Perhaps more importantly, we have found that, well before a selection equilibrium has been reached, the fitness advantages of new mutants become so small that their fate is primarily determined by drift even when such mutants have higher fitness than the current resident. (We provide more details on this phenomenon in the Appendix section for this chapter.) As evolution proceeds, there are fewer mutants with positive fitness generated and more mutants with neutral fitness generated. Since the fitness advantage of favored mutants is also becoming smaller as evolution proceeds, we see evolution dominated by drift, with some type of drift-selection balance over very large time spans. The further characterization of this drift-selection balance awaits future research.

Mortality Evolution Under Different Demographic Selection

The outcome of repeated introductions of new mutants on the evolution of mortality from an initially Gompertz pattern is shown in Figure 3-2. Over evolutionary time, early mortality declines and late mortality increases, although the exponential pattern of mortality increase is lost at advanced ages and the

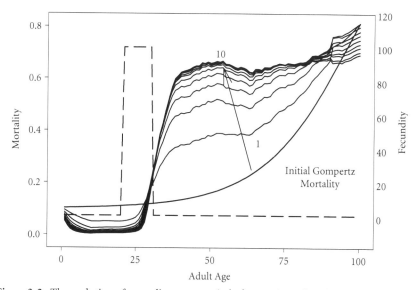

Figure 3-2. The evolution of mortality over a period of 10,000 introduced mutants. The arrow shows the direction of change in the mortality curve from the initial Gompertz pattern to a pattern with a pronounced plateau. The numbers 1 and 10 are next to curves that show the progress of evolution after 1,000 and 10,000 introduced mutants, respectively. The dashed curve represents the age-specific fertility pattern, which is assumed constant during the course of this simulation. Fecundity is 1 at all adult ages except 21–30, where it is set to 100.

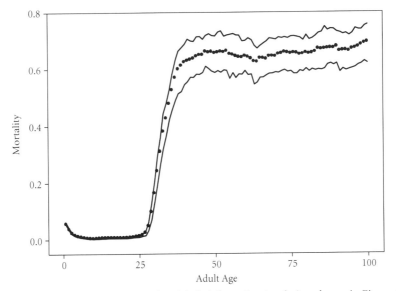

Figure 3-3. The average final mortality (circles) from the simulation shown in Figure 3-2 along with a 96% confidence interval (lines).

simulated patterns resemble a plateau. This plateau sets in at ages just beyond the peak age of reproduction, which in this example is 30 days.

The plotted lines in Figure 3-2 show the progression of mortality evolution averaged over the 100 different mutant orders. The average final evolved state and a 96% confidence interval about it are shown in Figure 3-3. As one might expect, the confidence interval is rather narrow around the averages at early ages, but it gets larger at late ages. This reflects the weakening of natural selection at late ages and the increasing influence of random genetic drift.

We kept track of the fitness of the resident population and show its change over time in Figure 3-4. Over the evolutionary time period of these simulations, there is a nearly monotonic increase in fitness, with the largest changes happening during the first 2,000–3,000 mutant introductions.

If the peak fecundity is moved to later ages, then the evolution of mortality should show a postponement of the age of onset of the plateau. This simple prediction follows from the fact that the delay in peak reproduction will make changes in survival relatively more important than they are when reproduction peaks at very young ages. We explored this prediction by altering the conditions of the simulation in Figure 3-2. We kept everything the same except the peak fertility, which was shifted from ages 41–50 rather than 21–30. The results of this case of simulated evolution, depicted in Figure 3-5, are similar to those seen in Figure 3-2, except that the plateau is delayed about 20 days—corresponding to the delay in peak reproduction. Thus our simple, if you will "intuitive," prediction is confirmed.

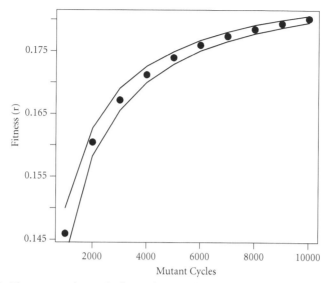

Figure 3-4. The average change in fitness (circles) for the simulation shown in Figure 3-2 along with a 96% confidence interval (lines).

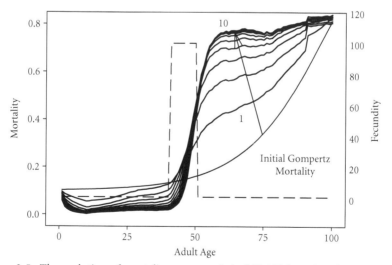

Figure 3-5. The evolution of mortality over a period of 10,000 introduced mutants. The arrow shows the direction of change in the mortality curve from the initial Gompertz to a pattern with a pronounced plateau. The numbers 1 and 10 are next to curves that show the progress of evolution after 1,000 and 10,000 introduced mutants, respectively. The dashed curve represents the age-specific fertility pattern, which is assumed constant. Fecundity is 1 at all ages except 41–50, where it is set to 100.

FECUNDITY EVOLUTION IN HAMILTONIAN THEORY

The evolutionary theory of late life based on the force of natural selection can also be applied to the evolution of age-specific fecundity. Like mortality, the age-specific force of natural selection acting on fecundity, $s'(x)$, has a scaling effect:

$$s'(x) = e^{-rx} l(x).$$

(3-5)

All the variables in Equation 3-5 have the same definitions as those in Equation 3-1. The force of natural selection acting on fecundity declines with age if population growth is not negative (Hamilton 1966; Charlesworth 1980, 1994). The probability of survival to age x directly affects the force of natural selection on fecundity at that age. According to this theory, $s'(x)$ will converge on zero after the last age of survival in the population's evolutionary history.

Hamilton's forces of natural selection acting on mortality and fecundity are similar in their effects, and thus will shape both age-specific mortality and fecundity within populations in a comparable manner. Therefore, the evolutionary theory of late life also predicts that late-life fecundity will roughly plateau at ages greater than the age at which $s'(x)$ declines to zero.

We can again illustrate this evolutionary inference using numerical simulations based on conventional age-structured population genetics. Our computer simulations had populations evolving with recurrent mutations to explore how age-specific fecundity is molded by natural selection. We assumed that survival followed the Gompertz equation and that environmental variation affected female fecundity such that a female of age i would have fecundity equal to $F_i = f_i + c f_i Z$, where, $Z \sim N(0,1)$. In a constant environment, fitness in an age-structured population is found from the solution, r_0, to the Lotka equation, $1 = \sum_{i=1}^{d} e^{-r_0 i} l_i f_i$, where d is the total number of age classes (Charlesworth 1994). Fitness of new mutant genotypes in a variable environment was determined from a stochastic growth rate parameter, $w = r_0 - \dfrac{c^2}{2T_0^2}$, where T_0 is the mean generation time (Tuljapurkar 1990, Eq. 15.2.1).

Random genetic drift can affect the fate of weakly beneficial or deleterious mutants. We modeled this by using the fitness of the resident and novel mutant genotypes to determine the probability of fixation from Ewens (1979, Eq. 3.28). A uniform random number was then chosen to simulate this fixation event.

In these particular simulations, the mutant fecundity schedules all exhibit antagonistic pleiotropy. Thus, a mutant was assumed to produce a stretch of 10 consecutive days of elevated fecundity and 10 consecutive days of depressed fecundity relative to the resident. The onset of elevated fecundity was chosen at random from the 100 possible age classes and similarly, but independently, for depressed fecundity. If the current resident's fecundity at day i was f_i, then a mutant's fecundity would be elevated to $f_i + (f_{max} - f_i)U$, where f_{max} is the maximum allowable fecundity set to 100 in these simulations, and U is a uniform random number between 0 and 1. Fecundity was depressed by $f_i - f_iU$. Each simulation run required the generation of 10,000 mutants. Their order was also shuffled, as was done previously with the mortality mutants.

The average results of 100 simulations show that fecundity evolves to a maximum level at young ages (Figure 3-6), but then declines rapidly and reaches a more or less constant value at about age 25–30 and thereafter. Thus, the force of natural selection becomes so weak at later ages that these ages eventually evolve an absence of differences, making fecundity plateau.

In Chapter 9, we discuss in more detail the catastrophic decline in fecundity that precedes death, which we call the *death spiral*. The death spiral will obscure the type of pattern predicted in Figure 3-6. To illustrate this effect, we have simulated the death process in cohorts of 2,000 females and computed the age-specific fecundity when it is genetically determined by the final curve in Figure 3-6. In addition, we have added the effects of a death spiral to these mean values using parameter estimates for this phenomenon taken from our work

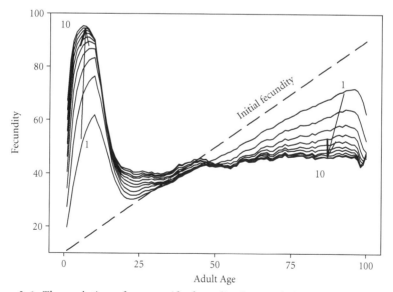

Figure 3-6. The evolution of age-specific fecundity in populations exposed to 10,000 mutants with antagonistic effects on age-specific fecundity characters. Initial survival probabilities were obtained from the same Gompertz mortality function used in Figures 3-5 and 3-2. Initial fecundity was assumed to increase with age. Similar results are obtained if fecundity is simply constant with age. The arrows show the direction of change in the fecundity curve. The numbers 1 and 10 are next to curves that show the progress of evolution after 1,000 and 10,000 introduced mutants, respectively.

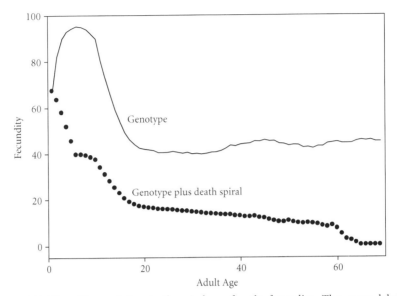

Figure 3-7. The effects of the death spiral on female fecundity. The curve labeled "Genotype" is the final fecundity schedule reached in the simulation shown in Figure 3-6. One hundred cohorts of 2,000 females were created. For each cohort, the ages of death were simulated using the Gompertz survival schedule used in the simulation in Figure 3-6. The mean fecundity at each age for each cohort was computed using the "Genotype" fecundity curve shown above and the model of female fecundity developed in Chapter 9. The slope of female fecundity in the death spiral was set to –0.2, which is similar to estimates for *Drosophila* in Mueller et al. (2007). The "Genotype plus death spiral" curve is the mean of the 100 simulated cohorts that incorporated the death spiral.

with *Drosophila* (Mueller et al. 2007). The results of applying these modifications to the results of Figure 3-6, as presented in Figure 3-7, show that even though the underlying age-specific fecundity curve is flat in late life, the overall pattern suggests a continual decline with age due to the effects of an ever-increasing fraction of the population in the death spiral.

As with mortality evolution, we can also take the 100 different simulations and compute a mean age-specific fecundity at the end of the evolutionary cycle and a 96% confidence interval. Thus, while the small details vary with the order in which these mutants are introduced, the overall pattern of an early peak in fecundity and a broad late-life plateau are always seen (Figure 3-8). The width of the confidence interval increases substantially at ages where fecundity has leveled off, which is consistent with the declining strength of selection.

Finally, we show the mean fitness trajectory over these simulations with a 96% confidence interval (Figure 3-9). The greatest fitness gains are made by the first 2,000 introduced mutants.

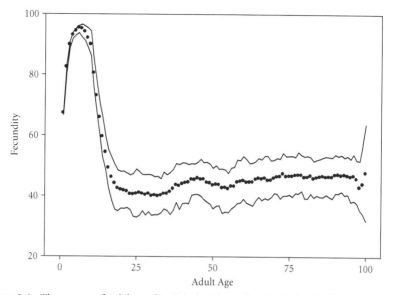

Figure 3-8. The average final fecundity (circles) from the simulation in Figure 3-6 along with a 96% confidence interval (lines).

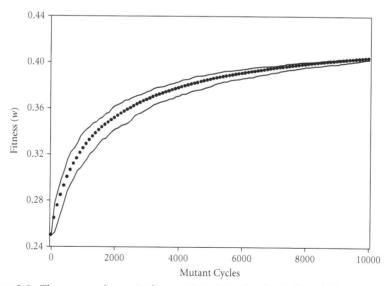

Figure 3-9. The average change in fitness (circles) for the simulation of Figure 3-6 along with a 96% confidence interval (lines).

JOINT EVOLUTION OF MORTALITY AND FECUNDITY

It may be more realistic to let new mutants have effects on both survival and fecundity. We have followed the simulated evolution of both traits under a model of antagonistic pleiotropy. That is, each mutant had a beneficial effect on either mortality or fecundity and a deleterious effect on the alternative trait. Both fecundity and mortality responded as they had previously (Figure 3-10), as did the population mean fitness (Figure 3-11).

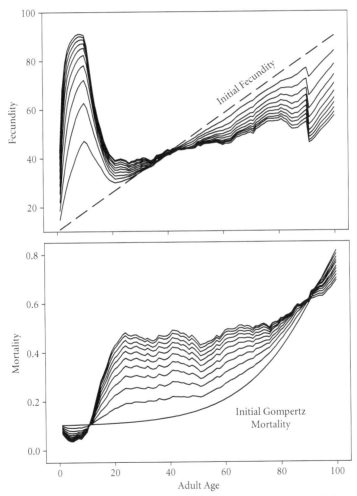

Figure 3-10. The evolution of age-specific fecundity and mortality in populations exposed to 10,000 mutants with antagonistic effects on fecundity and mortality. Initial survival was the same Gompertz mortality used in Figures 3-5 and 3-2. Initial fecundity was assumed to increase with age, as in Figure 3-6. Each line shows the progression of evolution after the introduction of 1,000 mutants, as in Figures 3-5 and 3-6.

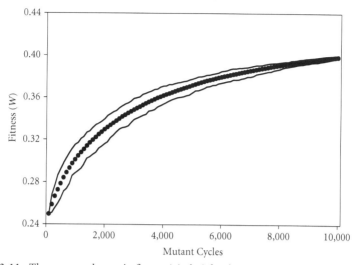

Figure 3-11. The average change in fitness (circles) for the simulation in Figure 3-10 along with a 96% confidence interval (lines).

OTHER MODELS OF LATE-LIFE PLATEAUS

Over the past 15 years, a number of additional models have been proposed to explain the observations of late-life mortality plateaus. Some have supported our general proposition that natural selection is the primary force in this evolution (e.g., Charlesworth 2001). Many other theories are nongenetic, or in some cases suggest novel schemes for the evolution of late-life plateaus. We review many of these additional theories in the Appendix section for this chapter.

We do not regard the simple numerical examples that we have given here as the last word on the subject of the evolutionary theory of late life. We expect our colleagues to produce a wide variety of interesting theories for the evolution of late life, together with instantiating analytical or simulation models. Indeed, we hope that they will do so.

But the ultimate arbiter of the validity of mathematical theories in all fields of science is whether they are corroborated or refuted by well-designed experiments. We are not naive about the relationship between experiments and theory. As experimentalists, we know that bad experimental design or execution can give results that may be incorrectly interpreted as supporting or falsifying particular theories. But over the course of sustained and careful experimentation, particularly using powerful tools like experimental evolution (vid. Garland and Rose 2009), we believe that the relative value of biological theories can be evaluated empirically. Thus, while a mathematical theory may be beautiful to contemplate, in the end its fate should depend more on a collection of ugly, but obdurate, experimental facts.

Our chief purpose here is to show that simple population genetic models *can* generate plateaus in later adult life, plateaus in both age-specific mortality and age-specific fecundity. Furthermore, we contend, these plateaus arise naturally from basic features of the sensitivity of natural selection to age-specific genetic effects. And the broad features of such sensitivity are captured reasonably well by Hamilton's twin forces of natural selection.

CONCLUSION: HAMILTON'S THEORY PREDICTS THE EXISTENCE OF MORTALITY AND FECUNDITY PLATEAUS

Although Hamilton's original insights were used to deduce a link between aging and natural selection, we have shown in this chapter that they also can be used to predict plateaued mortality and fecundity patterns during late life. At such advanced ages, selection may no longer distinguish between genetic effects among late ages; thus, later-life history can evolve toward plateaus with high mortality rates and low fecundity at these later ages. This theory is amenable to experimentation. In particular, it has the corollary that when the strength of age-specific selection is manipulated, the age at which plateaus are observed to start should evolve. We describe tests of this theory in the next chapter.

Late-Life Mortality and Fecundity Plateaus Evolve

In Drosophila laboratory evolution, mortality and fecundity plateaus evolve in the manner predicted by Hamiltonian theory. These "strong-inference" experiments provide corroboration for the Hamiltonian interpretation of late life.

EXPERIMENTAL EVOLUTION AS A TECHNIQUE FOR TESTING HAMILTONIAN AGING THEORY

Experimental evolution is a powerful technique for testing evolutionary theories of all types (vid. Garland and Rose 2009). Indeed, one of its earlier and most successful applications was in tests of Hamilton's original use of the forces of natural selection to explain the evolution of aging (Rose and Charlesworth 1980; Rose et al. 2007). Compared to the use of genetic variances and covariances, experimental evolution has been a consistently more reliable technique for performing strong-inference tests of Hamiltonian theory (Platt 1966; Rauser et al. 2009). Genetic variances and covariances among life-history characters are subject to tricky inbreeding and genotype-by-environment interactions (Rose 1991). While it has been found that experimental evolution is also subject to these problems, it has been possible to sort these artifacts out with further experiments (e.g., Leroi et al. 1994a,b). Similar progress with experimental tests focused on variance components has proven considerably more difficult (vid. Shaw et al. 1999).

The key experimental trick used to test the Hamiltonian explanation of aging is to postpone the *first* day of reproduction in outbred laboratory populations and then to sustain that regime for multiple generations of experimental evolution. This is done by keeping adult flies alive for some time before they are allowed to contribute offspring to the next generation. This can be achieved by discarding any eggs that they lay until they have reached the age allowed for reproduction, which can be as late as 10 weeks from emergence of the larva. Note that this procedure does not require that the fruit flies be kept virgin; mating can be allowed, just not *successful* reproduction. This regime is expected to lead to the evolution of relatively later aging. Wattiaux (1968) and Rose and Charlesworth (1980, 1981) found evidence of enhanced later-age fertility and longevity in *Drosophila* populations cultured with later ages of first reproduction without replication of selected or control populations.

Properly replicated experiments using this experimental approach were not performed until the 1980s, particularly by Rose (1984b) and Luckinbill et al. (1984). Rose (1984b) analyzed longevity and fecundity differences among three populations selected for earlier reproduction and three populations selected for increasingly later first ages of reproduction. These early- and late-reproducing populations were derived from the same outbred laboratory population of *D. melanogaster*, but had been separated and selected for their relative ages of reproduction for more than 15 generations at the time they were employed in the first assays. Significant differences were observed in longevity between the early and late reproducers, with the late reproducers having increased mean longevity (Figure 4-1). Luckinbill et al. (1984) found essentially the same results, further demonstrating that selection on first age of reproduction can alter longevity in ways consistent with the Hamiltonian explanation of the evolution of aging. Experiments using the method of delayed first reproduction are now routine, often using fruit fly species of the genus *Drosophila,* but sometimes other species are used (e.g., Nagai et al. 1995; Reed and Bryant 2000).

EXPERIMENTAL STRATEGY OF CONTROLLING THE LAST AGE OF REPRODUCTION AND SURVIVAL

In Chapter 3, mortality-rate patterns were predicted to follow the pattern of the force of natural selection and to plateau sometime after the force of natural selection plateaus in late life (see also Mueller and Rose 1996; Rose and Mueller 2000;

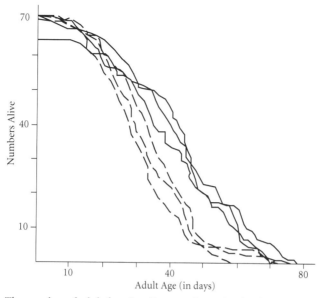

Figure 4-1. The number of adult females alive at each age for the three earlier-reproduced populations (dashed lines) and the three later-reproduced populations (solid lines). Later-reproduced populations demonstrated an increase in mean longevity (42.81 days) compared to the earlier-reproduced populations (33.28 days) after just 15 generations of selection for progressively postponed reproduction.
SOURCE: Rose 1984b, p. 1006, Fig. 1.

Charlesworth 2001). The correspondence between the start of the plateau in the force of natural selection and the onset of mortality-rate plateaus is not expected to be exact, however, because beneficial gene effects that continue from early to late ages will sustain survival somewhat longer than the last age of reproduction in the evolutionary history of the population. Nevertheless, the age when mortality-rate acceleration stops, or slows, should evolve in accordance with large changes in the age at which the force of natural selection hits zero. Therefore, the mortality-rate pattern in late life is predicted to evolutionarily follow the pattern of prior selection on the population's last age of reproduction. If experimental populations with abundant genetic variation, whose last age of reproduction has been controlled in a consistent manner for numerous generations, do not conform to this pattern, then the evolutionary theory for late-life mortality based on the force of natural selection would be falsified. We will now review experimental studies that were designed to test these predictions.

All stocks used in the experiments to be discussed in this chapter were ultimately derived from a sample of the Amherst, Massachusetts, Ives population (e.g., Ives 1970) that was collected in 1975 and has been cultured at moderate to large population sizes ever since. Individual populations have been subjected to a series of selection regimes, as indicated in Figure 4-2 (Rose 1984b; Chippindale et al. 1994). Each of four distinct types of stocks differs in its age of last reproduction, and each stock in turn consists of five outbred replicate populations. What we mean by "age of last reproduction" in these stocks is that individuals in these populations are allowed to mate freely and lay eggs on all days leading up to their last age of reproduction; however, the way in which these populations are cultured only allows eggs that are laid shortly before the last age of reproduction to contribute to the next generation. Therefore, the last age of reproduction marks the end of a brief window of successful reproduction. The four stocks are B_{1-5}, O_{1-5}, CO_{1-5}, and ACO_{1-5} (subscripts 1–5 indicate the five replicated populations within each stock). The ACO and B populations have an early age of last reproduction (9 and 14 days from egg, respectively), the CO populations have an intermediate last age of reproduction (28 days), and the O populations have a late last age of reproduction (70 days). These populations have each been maintained for more than 100 generations under their distinctive demographic regime at effective population sizes ≥1,000, and they are known to be highly polymorphic genetically (Rose et al. 2004). Together, these populations define a spectrum of selection on the age of reproduction, and thus a spectrum of patterns for the age-specific force of natural selection.

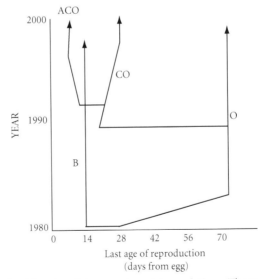

Figure 4-2. Selection histories of the experimental populations. The ancestral population was the IV population, sampled from nature in 1975, which was used as the ancestor of the five B and five O populations in 1980. In 1989, the five CO populations were derived from the five individual O populations, and the five ACO populations were in turn derived from each of the CO populations in 1991.

TESTING WHETHER MORTALITY PLATEAUS SHIFT WITH THE LAST AGE OF REPRODUCTION

We tested the Hamiltonian mortality-plateau prediction that the onset of mortality plateaus should evolve in accordance with the last age of reproduction in the population's evolutionary history, using the B, O, ACO, and CO *Drosophila* populations described above (see Figure 4-2). Survivorship assays employing these stocks already indicated that the last age of reproduction was positively correlated to the lifespan. That is, the populations with the earliest last ages of reproduction (B and ACO populations) had shorter lifespans than populations with later last ages of reproduction (CO and O populations). [Data on the lifespans of these laboratory populations are compiled in Rose et al. (2002, 2004)]. However, these average lifespan patterns by themselves do not indicate the timing or nature of the mortality-rate plateaus of these populations.

In order to determine whether the timing of mortality-rate plateaus evolves according to the last age of reproduction in a population's evolutionary history, mortality-rate comparisons between populations that were evolutionarily distinct with respect to their ages of last reproduction had to be done. Therefore, we performed two mortality-rate comparisons using the populations described above to provide independent tests of the evolutionary theory for late-life mortality plateaus. These were the comparisons of the B with the O populations and the ACO with the CO populations. Specifically, these comparisons allowed us to test the prediction concerning the effects of last age of reproduction on the start of mortality-rate plateaus.

The B and O populations share a common ancestor but have long had a 56-day difference in their last age of reproduction. That is, they had evolved separately for more than 17 years (450 B-generations) at the time that we estimated the age-specific mortality rates presented here. The mortality data of the B and O populations were fit to two-stage Gompertz equations by maximum likelihood techniques, allowing, but not assuming, a late-life mortality rate plateau (see "Estimation of Mortality Rate Plateaus" in the Appendix). This model fitting was not performed in order to support the Gompertz model, but rather merely to infer mortality-rate patterns by an objective procedure.

The ACO populations were derived directly from the CO populations, as shown in Figure 4-2. The ACO populations had a last age of reproduction of 9 days, while the CO populations had a last age of reproduction averaging about 28 days in the period before the present experiments. These populations were compared using a paired-comparison test because each ACO population was derived from the CO population having the same numeric subscript.

In these two large-scale and independent comparisons, we collected mortality data for each replicate population for both males and females, starting at the 9th or 10th day of age from egg until all flies in the cohort had died (see Rose et al. 2002 for experimental details). We tried to avert problems in interpreting the mortality data by controlling for effects arising from varying density and small population sizes. For example, males and females were housed together, density was kept roughly constant throughout each assay (cf. Carey et al. 1993; Nusbaum et al. 1993; Graves and Mueller 1993, 1995; Curtsinger 1995a,b; Khazaeli et al. 1995a, 1996), and high cohort sizes (at least 2,000 individuals per replicate) were used to reduce sampling variance in our estimations of mortality rates (cf. Pletcher 1999; Promislow et al. 1999) (see Tables 4-1 and 4-2 for sample sizes).

Mortality data from the B–O and ACO–CO populations were fit to a two-stage Gompertz model using maximum likelihood techniques, as described in the Appendix. This model allows, but does not assume, a late-life mortality rate plateau, although we observed plateaus in mortality rates in all of our replicate populations at later ages. The importance of this two-stage model is that it allowed us to estimate the approximate age at which mortality rates started to plateau, or the breakday between the two stages of the model, within each population. This, in turn, permitted us to test the Hamiltonian mortality-plateau prediction that the onset of mortality plateaus should evolve in accordance with the last age of reproduction in the population's evolutionary history.

Table 4-1. Results from a test of the evolutionary theory for late-life mortality using comparison between the early-reproducing B populations and the late-reproducing O populations with respect to onset of mortality-rate plateaus. A and α are from the Gompertz equation.

	Males			Females		
	B	O		B	O	
Sample size	4,867	8,855		5143	10,037	
Breakday	23.6	58.0	***	24.0	68.4	***
Plateau mortality rate	0.338	0.161	***	0.240	0.195	***
A	0.00339	0.00124	*	0.00542	0.00307	
α	0.198	0.0711	***	0.173	0.0577	***
Mean longevity	20.6	52.3	***	20.8	48.2	***

$* \, p < 0.1;$ $*** \, p < 0.01.$

Table 4-2. Results from an independent test of the evolutionary theory for late-life mortality using a comparison between the earlier-reproducing ACO populations and the later-reproducing CO populations with respect to onset of mortality-rate plateau. Because each ACO population derives from a single CO population, paired-difference *t*-tests were used to test for significant differences between characters.

	Males			Females		
	ACO	CO		ACO	CO	
Sample size	12,444	11,987		14,084	12,361	
Breakday	42.6	58.6	***	40.6	57.0	***
Plateau mortality rate	0.363	0.286	**	0.520	0.330	***
A	0.00500	0.00156	***	0.00710	0.00465	**
α	0.106	0.0813	***	0.105	0.0644	***
Mean longevity	26.2	44.2	***	23.5	37.2	***

** $p < 0.05$; *** $p < 0.01$.

MORTALITY PLATEAUS EVOLUTIONARILY SHIFT WITH THE LAST AGE OF REPRODUCTION

These two independent comparisons between laboratory-evolved populations selected for different last ages of reproduction tested the predictions made by the Hamiltonian evolutionary theory for late-life mortality and our computer simulations described in Chapter 3. Specifically, we predicted that the five later-reproducing O populations would have a later onset of mortality-rate plateaus compared to the earlier-reproducing B populations. Similarly, in the pairwise comparison between the CO and ACO populations, we predicted a later onset of mortality-rate plateaus in the five later-reproducing COs compared to the five ACOs. This is exactly what we found (see Figures 4-3 and 4-4 and Table 4-1). Our experimental predictions were confirmed and our results fully corroborated the Hamiltonian mortality-plateau prediction.

Notably, both experiments described above could have refuted the evolutionary theory if there had been no difference between populations in the breakday of their mortality-rate plateaus, after long maintenance of very different terminal ages for reproduction, or if the difference between these breakdays had been in the opposite direction from the difference in the last day of reproduction.

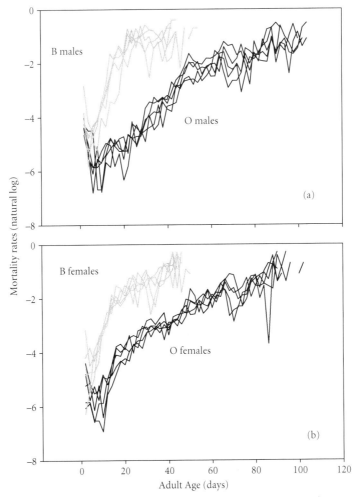

Figure 4-3. Two-day mortality rates for 10 cohorts sampled from the B and O populations. B populations (early last reproduction) are shown as gray lines, and O populations (late last reproduction) are shown as black lines. Occasional regions are missing because mortality rates of zero cannot be properly interpreted on a logarithmic scale. The ages at which late-life mortality plateaued in the O populations (male: 58.0; female: 68.4) were significantly greater than the ages at which mortality plateaued in the B populations (male: 23.6; female: 24.0). (a) Male mortality. (b) Female mortality.

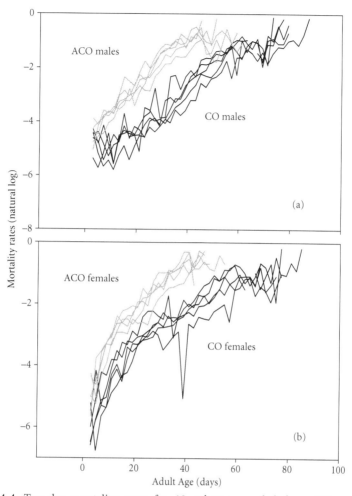

Figure 4-4. Two-day mortality rates for 10 cohorts sampled from CO and ACO populations. In each case, ACO populations (selected for early-life fecundity) are shown by gray lines and CO populations (selected for midlife fecundity) are shown by black lines. Occasional regions are missing because mortality rates of zero cannot be properly interpreted on a logarithmic scale, except for one case, which was a result of experimental error. Late-life mortality plateaued later in the CO populations (male: 58.6; female: 57.0) compared to the ACO populations (male: 42.6; female: 40.6). The ACO–CO comparison is a completely independent test of Hamiltonian evolutionary theory from the B–O comparison in Figure 4-3. (a) Male mortality. (b) Female mortality.

DETERMINING WHETHER FECUNDITY PLATEAUS AT LATE AGES

In Chapter 3, we discussed how the Hamiltonian theory based on the declining force of natural selection with age can just as easily be applied to fecundity as to mortality. The greatest difference between these two characters is that the force of natural selection acting on fecundity should decline with age until the last age of *survival* in the environment in which a population has evolved, rather than the last age of reproduction, which is the case with the evolution of mortality (Hamilton 1966). The force of natural selection acts on age-specific fecundity scales according to $s'(x) = e^{-rx} l_x$, where x is the age of a genetic effect on fecundity, r is the Malthusian parameter for the population, and l_x is survivorship to age x (Figure 4-5). After the last age at which individuals survive in the population's evolutionary history (say d, which is not necessarily the last age of cohort survival under protected conditions), $s'(x)$ converges on and remains at zero thereafter.

According to this evolutionary theory, the evolution of fecundity should echo its age-specific force of natural selection. That is, fecundity should decline in midlife and plateau at very late ages, in a fashion analogous to mortality rates. However, as with mortality, it may not be possible to detect these plateaus in female fecundity unless very large cohorts are examined. If we examine age-specific fecundity in a variety of organisms, some general patterns emerge

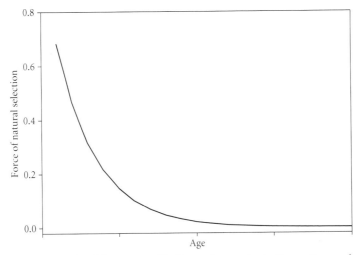

Figure 4-5. An example of the age-specific force of natural selection acting on fecundity. Even in organisms that reproduce indefinitely, the strength of selection may be so weak in late life that random genetic drift is the primary determinant of the frequency dynamics of alleles that differ only with respect to their effects sufficiently late in adult life.

(Figure 4-6). It is important to note here that when we refer to fecundity, we are talking about the average age-specific fecundity within a population, not individual female fecundity patterns. As we will describe in detail in Chapter 9, the relationship between average fecundity and individual fecundity is greatly complicated by the effect of dying on each type of fecundity.

Only the data from the flatworm, *Dugesia lugubris*, suggest a fecundity plateau in late life. However, all four species show an increase in fecundity following sexual maturity until it peaks sometime in early life or midlife, followed by a decline at later ages. Broadly speaking, we would suggest that many organisms show a unimodal age-specific fecundity curve that may either decline steadily to a low value or show some type of plateau at late ages, leaving aside seasonal reproduction patterns. Our simulations (Chapter 3) of the evolution of fecundity support a general pattern of decline from a peak in early life to a plateau at later ages (Figure 4-7).

To our knowledge, ours was the first laboratory to demonstrate empirically that fecundity within a population peaks during midlife and then declines to a low level and plateaus at late ages, as predicted by the fecundity model in Figure 4-7. We described in Chapter 2 our observations that fecundity indeed plateaus at late ages in several independent *Drosophila* populations (Rauser

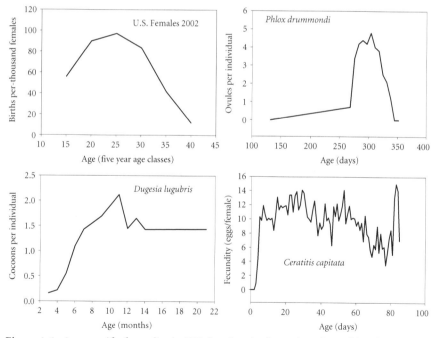

Figure 4-6. Age-specific fecundity in U.S. females: the flowering plant *Phlox drummondi*, the flatworm *Dugesia lugubris*, and the Mediterranean fruit fly *Ceratitis capitata*.

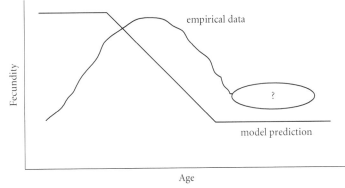

Figure 4-7. The expected shape of age-specific fecundity within a population. The evolutionary model developed in Chapter 3 predicts the curve labeled "model prediction." The general pattern from many organisms is labeled "empirical data." The question as to whether data collected sufficiently late in adult life will exhibit a plateau pattern is indicated by the question mark.

et al. 2003, 2006b). Although the late-life plateau in fecundity was not always distinct, we always observed a significant slowing in the decline in fecundity at late ages and used a variety of statistical tests to determine whether fecundity at late ages was significantly different from zero (see Rauser et al. 2003, 2006b). It is likely that previous experimental examinations of population fecundity did not reveal plateaus in fecundity at late ages because of small starting sample sizes. Our experiments consistently employed thousands of flies per cohort.

TESTING WHETHER FECUNDITY PLATEAUS EVOLUTIONARILY SHIFT WITH THE LAST AGE OF SURVIVAL

Analogously to the experimental tests of the Hamiltonian theory that we performed with regard to mortality described above, we tested the Hamiltonian fecundity-plateau prediction that the onset of fecundity plateaus in a population should evolve in accordance with the last age of survival in the population's evolutionary history. This prediction was tested in the ACO and CO *Drosophila* populations described above (see Figure 4-2), which are evolutionarily distinct with respect to their last ages of survival.

The difference in age of reproduction between the ACO and CO populations resulted in late-life mortality-rate plateaus that started at a significantly greater age in the CO populations relative to the ACO populations (Rose et al. 2002), as was predicted by the Hamiltonian theory. The difference in the age of reproduction between these populations is positively correlated with the age of last survival because of the way these populations are maintained and cultured. That is, once these populations have been allowed their successful day of reproduction (age 9 days in the ACO populations and 28 days in the CO populations), they are discarded, which hence also defines their last age of survival. Therefore, this difference in age of reproduction corresponds to the ages at which the force of natural selection acting on fecundity declines to zero and plateaus earlier in the ACO populations relative to the CO populations. Together, these 10 populations provide a platform with which to test the evolutionary theory of late life, based on the force of natural selection, as it applies to fecundity.

During the pairwise comparisons between each replicate ACO population and the corresponding CO population, four adult females were housed with four adult males in vials containing enough yeast so that mating and nutrition were not limiting factors for fecundity. Flies were also recombined between vials as flies died to forestall any age-specific density effects. The fecundity within these cohorts was determined daily until all flies had died. All assays started with 3,200 females per replicate population and as many males (see Rauser et al. 2006b for experimental details).

Fecundity data from each of the five ACO and five CO populations in the pairwise comparison were independently fit to a two-stage linear model, analogous to the two-stage model fit to our mortality data and described in detail in the Appendix, to test whether fecundity plateaus evolve according to Hamiltonian evolutionary theory. Specifically, the age of onset of the late-life fecundity plateau for a population, or the breakday, was estimated from the two-stage model and then used to test whether late-life fecundity plateaus evolve according to the age at which the force of natural selection acts on fecundity plateaus.

Population estimates of age-specific fecundity are complicated by the existence of flies that are about to die and those that are not. We have shown that females about to die show a rapid decline in fecundity no matter how old they are (Rauser et al. 2005b; Mueller et al. 2007). Nevertheless, the techniques used here can still reliably infer the onset of the fecundity plateau (see Mueller et al. 2007 for more details; also see further discussions below, particularly in Chapter 9).

FECUNDITY PLATEAUS EVOLUTIONARILY SHIFT WITH THE LAST AGE OF SURVIVAL

This pairwise comparison between laboratory-evolved populations selected for different last ages of reproduction, and consequently different last ages of survival, tested the predictions made by the Hamiltonian evolutionary theory for late-life fecundity. We specifically predicted that the later-reproducing CO populations would have a later onset of fecundity-rate plateaus compared to the earlier-reproducing ACO populations. This is exactly what we observed (see Figure 4-8 and Table 4-3). We found an average pairwise difference in the onset of the late-life fecundity plateaus of 13.80 days between the two selection

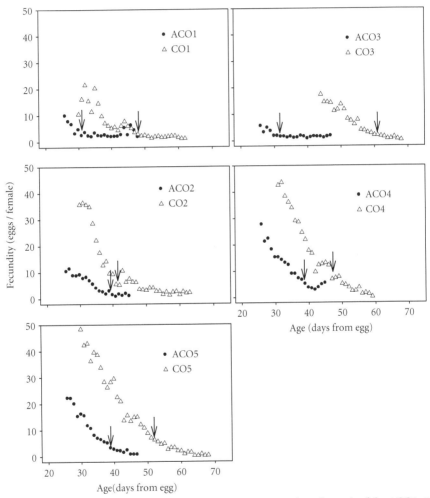

Figure 4-8. Mean mid- and late-life fecundity as a function of age for each of the ACO1–5 (early reproducing) and CO1–5 (late reproducing) populations. Fecundity was measured during the ACO*i* and CO*i* pairwise comparisons. A two-stage linear model was fit to each population independently (see the Appendix for details). For all 10 populations, plateau height was significantly greater than zero. Late-life fecundity plateaued later in the CO populations (49.86 days) compared to the ACO populations (36.06 days), as predicted by Hamiltonian theory ($p < 0.0001$). The arrows indicate the start of the fecundity plateau.

Table 4-3. Parameter estimates from the two-stage linear model fitted to mid- and late-life fecundity data from the earlier-reproducing ACO populations and the later-reproducing CO populations. The height of the fecundity plateau was computed from Equation A4-2, and the estimated height was significantly different from zero ($p < 0.05$ for each population).

Population	First-stage y-int (c_1)	First-stage slope (c_2)	Breakday (*fbd*)	Plateau height (eggs/female/day)
ACO_1	48.11	−1.49	30.52	2.50
ACO_2	30.61	−0.75	39.44	1.22
ACO_3	22.97	−0.69	31.62	1.21
ACO_4	67.44	−1.66	38.24	3.98
ACO_5	63.99	−1.61	38.44	2.16
CO_1	40.54	−0.80	48.55	1.80
CO_2	137.41	−3.26	40.67	4.86
CO_3	55.74	−0.90	60.43	1.32
CO_4	121.26	−2.51	46.30	5.27
CO_5	101.66	−1.89	51.66	3.81

NOTE: Parameter estimates for c_1, c_2, and *fbd* were all significantly different from zero; $p < 0.001$.

regimes (Figure 4-9 and Table 4-4). As with mortality, our results for fecundity fully corroborated the Hamiltonian late-life prediction. The plateaus in fecundity evolved according to the age at which the force of natural selection acting on fecundity declined to zero.

The pairwise comparison between the two replicated sets of populations long having different last ages of survival in their evolutionary histories would not have supported the evolutionary theory for late life as described by Hamilton if there had been no difference between the populations in the onset of their fecundity plateaus (breakday). Furthermore, the theory would not have been supported if the onset of these fecundity plateaus had been in the opposite direction from the difference in the last age of survival in their respective evolutionary histories. However, that was not the case.

Most evolutionary theories suggest a rapid rise in age-specific fecundity at early ages followed by a long decline after some peak value. Our interpretation of the evolutionary theory of late life, based on the decline in the force of natural selection, was that population fecundity will plateau at very late ages, like age-specific mortality rates (Rauser et al. 2003). We made this prediction because

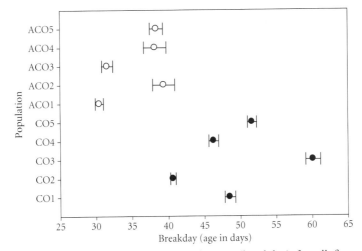

Figure 4-9. Late-life fecundity plateau age of onset (breakday) for all five pairwise comparisons of CO and ACO populations. The fecundity plateau started significantly later in the later-reproducing CO populations compared to the earlier-reproducing ACO populations ($p < 0.0001$). The breakday and 95% confidence intervals were estimated for each population from the two-stage linear model using a nonlinear least squares regression function.

Table 4-4. Results from the test comparing the fecundity-model parameters of the earlier-reproducing ACO populations and the later-reproducing CO populations. Plateau height was computed from Equation A4-2. The x,y-values used in the regression were from 100 vials (400 females) randomly sampled daily from an initial population size of 4,000 vials (16,000 females).

	Population		
	ACO	**CO**	
Sample size (x,y-values)	6,540	12,873	
First-stage y-int	43.58	88.31	*
First-stage slope	−1.14	−1.80	
Fecundity breakday	36.06	49.86	***
Plateau height (eggs/female/day)	2.21	3.41	

*** $p < 0.0001$; * $p < 0.05$.

the force of natural selection acting on age-specific fecundity asymptotically falls to such a low level that it can no longer distinguish fitness differences in fecundity at different ages. Our experimental work supports this interpretation. We found that the decline in fecundity greatly slows, or plateaus, in 10 independent populations at some number of eggs laid per day greater than zero. Furthermore, we found that fecundity plateaus evolve according to the age of last survival in these populations' evolutionary histories. These results corroborate the basic evolutionary theory of late life and its prediction that fecundity, just like mortality rates, should plateau some time after the age-specific force of natural selection acting on fecundity itself plateaus.

EXPERIMENTAL EVOLUTION SUPPORTS THE HAMILTONIAN THEORY OF LATE LIFE

Over the past decades, our laboratory has tested the predictions of Hamilton's evolutionary theory based on the age-specific decline in the forces of natural selection with respect to both mortality and fecundity using experimental evolution techniques in numerous *Drosophila* cohorts. Hamilton's classic theory (1966) predicts that the force of natural selection acting on these characters should decline to zero at late ages, or to levels so insignificant that age-specific natural selection is essentially nonexistent at such late ages. The timing in the decline in the force of natural selection to zero is dependent on the last age of reproduction and survival in the population's evolutionary history for mortality and fecundity, respectively.

Therefore, by employing populations of flies that have long undergone selection for specific and different ages of last reproduction and survival, we were able to test the specific predictions of the Hamiltonian theory that apply to late life. For our experiments, our numerical calculations led us to the hypothesis that the onset of mortality-rate plateaus at late ages should evolve, or shift, in accordance with the last age of reproduction in that population's evolutionary history. Similarly, the onset of fecundity plateaus at late ages should evolve according to the last age of survival in that population's evolutionary history. This is precisely what we observed. In sum, all of our experiments have corroborated Hamiltonian theory, as we have observed the evolution of both mortality rate plateaus and fecundity plateaus at late ages as predicted by that theory. Next, we turn to the genetic mechanisms that might underlie these experimental evolutionary results.

Genetics of Late Life Involve Antagonistic Pleiotropy

Reverse evolution experiments implicate antagonistic pleiotropy in the evolution of both mortality and fecundity during late life. Hybridization experiments do not implicate mutation accumulation in the evolution of late life but do not necessarily preclude its involvement.

POPULATION GENETICS OF AGING AND LATE LIFE

The population genetics of aging without regard to late life have been theoretically developed and empirically tested, starting at around the mid-twentieth century. There are two main population genetic mechanisms that can explain aging: mutation accumulation and antagonistic pleiotropy (see Rose 1991). These two mechanisms can function separately or in concert, which means that they are not mutually exclusive. Furthermore, these same population genetic mechanisms can explain plateaus in age-specific mortality rates and the evolution of late life in general (Mueller and Rose 1996; Charlesworth 2001).

The ways in which mutation accumulation and antagonistic pleiotropy function in the evolution of aging and late life, and how these mechanisms can be empirically tested, will be described in turn. Mutation accumulation affects the evolution of aging and late life when alleles that are deleterious at later ages, but neutral at all earlier ages, accumulate by mutation pressure and genetic drift (Medawar 1952; Rose 1991; Charlesworth 1994, 2001). Such mutations are expected to be unique to each evolving population. They are also expected to be somewhat recessive on average, since that is usually the heterozygous effect of deleterious mutations (Simmons et al. 1978). Despite their deleterious effects, these mutations are able to persist in populations because they increase mortality rates only later in life, when the force of natural selection is relatively weak. These features of mutation accumulation are expected to produce hybrid vigor in experimental crosses of populations subject to mutation accumulation. However, it is important to note that not all alleles are expected to foster hybrid vigor with mutation accumulation, and mutation accumulation is not the only possible cause of hybrid vigor (Charlesworth and Hughes 1996). Nevertheless, the demonstration of hybrid vigor in crosses between populations influenced by mutation accumulation provides at least indirect support for the hypothesis of mutation accumulation as a genetic mechanism in the evolution aging and late life, as argued by Mueller (1987) and Rose et al. (2002).

Another genetic mechanism that may explain the evolution of aging and late life is antagonistic pleiotropy, specifically when alleles that are beneficial early in life are deleterious later in life (Williams 1957; Rose 1985; Charlesworth 1994). With mortality, for example, alleles that are deleterious and cause increased mortality rates late in life can persist within a population because these same alleles enhance another fitness-related trait, such as reproduction, earlier in life when the force of natural selection is much stronger. For life-history evolution, this genetic mechanism can be experimentally distinguished from mutation accumulation and genetic drift by subjecting long-established late-reproducing populations to an evolutionary reversion to much earlier ages of reproduction, an experimental protocol that has been of value in the study of the evolution of aging (e.g., Service et al. 1988). So long as this reverse selection (cf. Teotónio and Rose 2001) is imposed on large populations for a small number of generations, there is too little evolutionary time for mutation accumulation or genetic drift to act significantly. For mortality, switching to a selection regime with an earlier last age of reproduction for a short amount of time should lead to an earlier-onset age for late-life mortality rate plateaus if this genetic mechanism is active in the evolution of late life. This experimental design tests whether antagonistic pleiotropy is operating in the evolution of late life because selection for early reproduction will increase the frequency of alleles enhancing early-fitness characters, and those alleles with antagonistic pleiotropy between early and late ages will in turn increase mortality rates before the start of the plateau, causing an earlier plateau onset. Therefore, if a shift in the age of onset of mortality plateaus to earlier ages is observable in the populations reverted to earlier ages of reproduction for a small number of generations, then antagonistic pleiotropy can be inferred as a genetic mechanism underlying late-life mortality patterns.

POPULATIONS EMPLOYED IN OUR TESTS OF GENETIC MECHANISMS

The stocks used in the experimental tests of the population genetics of late life described in this chapter were ultimately derived from a sample of the Amherst, Massachusetts, Ives population (e.g., Ives 1970) described in Chapter 4 (see Figure 4-2 and Figure 5-1). Recall that each of the stocks differs in its age of last reproduction, which is controlled in the laboratory by the way in which the stocks are cultured. Furthermore, each of these stocks in turn consists of five outbred replicate populations (Rose 1984b; Chippindale et al. 1994). The four stocks, described earlier, are the B_{1-5}, O_{1-5}, CO_{1-5}, and ACO_{1-5}. The ACO and B populations have an early age of last reproduction (9 and 14 days from egg, respectively), the CO populations have an intermediate last age of reproduction (28 days), and the O populations have a late last age of reproduction (70 days). These populations had been maintained for more than 100 generations at population sizes $\geq 1,000$ at the time of the experiments described here. Together, these populations define a spectrum of selection on the age of reproduction, and thus a spectrum of patterns for the age-specific force of natural selection acting on mortality. As described in Chapter 4, the timing of the onset of mortality-rate plateaus in these populations corresponds positively with the last age of reproduction in the evolutionary history of the populations. That is, late-age plateaus in mortality occurred earliest in the ACO populations, followed by the B, CO, and O populations. For testing the population genetic theories of aging and late life, the B populations were employed in an experiment to test whether mutation accumulation contributes to the evolution of late-life mortality rate plateaus, and the O and CO populations were used to test the theory of antagonistic pleiotropy in the evolution of late-life mortality and late-life fecundity, respectively.

To properly test the role of antagonistic pleiotropy in the evolution of aging and late life, new stocks were created that originated from the O and CO stocks. To specifically test whether antagonistic pleiotropy influences the evolution of late-life mortality, the O populations were reverted to an earlier last age of reproduction (14 days) for only 24 generations prior to the experimental assays. This new stock was named NRO_{1-5} and each of the five NRO populations was derived from its respective O population (Figure 5-1). The NRO culture procedure was like that of

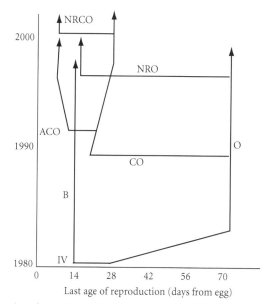

Figure 5-1. Selection histories of the reverse-selected populations. The five NRO populations were derived from the individual O populations in 1998, and four NRCO populations were derived from the corresponding CO populations in 2003 to test whether the population genetic mechanism antagonistic pleiotropy operates in the evolution of late life.

the O populations, except that flies were placed in cages at about 10 days from egg, followed by egg collection at 14 days, after feeding with yeast.

In order to test whether antagonistic pleiotropy contributes to the evolution of late-life fecundity, four NRCO populations were derived from their corresponding CO populations and subjected to selection for earlier reproduction using procedures similar to those routinely used with the ACO populations (see Figure 5-1). This reverse selection was imposed for only 24 generations, as with the NRO populations described above, after which time the experimental assays were performed.

TESTING WHETHER MUTATION ACCUMULATION ACTS AS A GENETIC MECHANISM IN THE EVOLUTION OF LATE-LIFE MORTALITY

In order to test whether late-life mortality does indeed reproducibly reflect some type of mutation accumulation, we generated 25 distinct outbred populations of *Drosophila melanogaster* by making all possible crosses of the five B populations, described in Chapter 4 (see Figure 4-2 and Figure 5-1). These populations were derived from a common ancestral population in February 1980, and since then have been kept on a 2-week culture regime with population sizes of approximately 1,000 individuals (Rose 1984b; Leroi et al. 1994a). Therefore, at the time the experiments described here were performed, about 18 years, or 465 generations, had elapsed since their founding. It is unlikely that a substantial number of new mutations affecting survival have arisen in these replicate B populations since their founding. Rather, it is more likely that each B population started with a large number of rare alleles that were deleterious in their effects on late-life survival while in nature, but were subsequently made neutral by laboratory culture. A fraction of these neutral alleles are expected to increase in frequency by random genetic drift. Furthermore, molecular studies of the five independent B populations have shown that they are genetically differentiated (Fleming et al. 1993).

The way in which the B populations are cultured actually creates conditions for mutation accumulation. Such an accumulation of mutations, however, depends on several factors, such as (i) the elimination of selection in late life, (ii) a finite population size, and (iii) the existence of late-acting deleterious alleles for the life-history characters we examine. The first two factors are part of the experimental design developed to culture these populations in the laboratory, while the third factor constitutes the biological hypothesis of mutation accumulation. The dynamic aspects of the process of mutation accumulation in the B populations are shown in Figure 5-2. This process presumes that multiple loci affect the

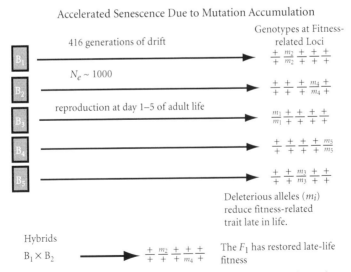

Figure 5-2. Alleles in the five independent B populations undergo independent rounds of genetic drift that can ultimately result in different sets of initially rare deleterious alleles rising to high frequency. When the different lines are crossed, the resulting hybrids are expected to be heterozygous and hence show improved late-life characters.

trait of interest. It further assumes that some existing deleterious alleles will rise to high frequency in each population and others will not. Furthermore, the particular alleles that rise to high frequency in one population are different than those alleles that rise to high frequency in the other populations.

We assume that most deleterious alleles that rise to high frequency by drift will be recessive or partially recessive. This conclusion follows from the simple population genetic considerations that suggest that, in the ancestral population, recessive deleterious alleles will be at a much higher equilibrium frequency than dominant alleles. Consequently, in the simplest case, which involves fixation of the deleterious alleles, the parental populations would show a depression in the same late-life characters that would be elevated in the F_1 hybrids created from crosses between any of the independent B populations (Figure 5-2). If on the other hand, fitness characters early and late in life are determined wholly by alleles with antagonistic effects, there should be little difference between the hybrids and parental populations for their late-life fitness characters when simultaneously compared.

However, it is important to note that there is nothing about the design of this experiment that guarantees or assumes that these deleterious alleles would be fixed after only 465 generations of drift. This is because, for neutral alleles at an initial frequency of p and with an effective population size of N, it will take on average $-4N(1 - p)\ln(1 - p)/p$ generations to fix the allele (assuming it is fixed, which will occur with probability p; Ewens 1979, p. 77). Therefore, in the B populations where $N_e \cong 1,000$, and assuming that $p = 0.05$, the average time to fixation would be 3,898 generations.

While 465 generations is not a sufficient amount of time for most initially rare neutral alleles to be fixed, there may be some that have risen to sufficiently high frequencies that late-life characters would be depressed. For instance, using the stationary distribution of neutral alleles, we can calculate the chance of finding neutral alleles in certain frequency ranges (Crow and Kimura 1970, p. 383). In the B populations, 4%–9% of the neutral alleles are expected to be at a frequency of 0.4 or greater (assuming $N_e = 1000$, and the initial frequencies are between 0.01 and 0.1). At final frequencies above 0.4, there would be sufficient numbers of homozygotes with deleterious effects at late ages, yet still neutral under B conditions, to reduce late-life fitness characters.

As a test of mutation accumulation underlying late-life mortality plateaus, we estimated the amount of hybrid vigor between the genetically divergent B laboratory selection lines. That is, every pairwise combination of the cross $B_i \times B_j$ (both i and j varying from 1 to 5) was performed, which resulted in 25 total crosses that included five parental (nonhybrid) and 20 hybrid fly cultures. The progeny from these crosses were assayed for mortality (for experimental details see Rose et al. 2002), similar to the mortality assays described in Chapter 4. Over 800 males and an equal number of females were assayed from each of the 25 resulting fly cultures, and mortality-rate estimations were done in the same manner for both hybrid and nonhybrid populations. Due to missing observations when collecting survival data, only 14 of the 20 hybrid crosses produced were included in the final analysis.

We found that the late-life mortality of the hybrids created from crosses between the five B populations exhibited no detectable difference from, or superiority to, the uncrossed cohorts sampled from the parental B populations for (i) overall longevity (Figure 5-3), (ii) onset of mortality-rate plateaus at late ages (Figure 5-4), or (iii) mean estimated mortality rate on the plateau (males: t-test, $p = 0.14$; females: t-test, $p = 0.46$). While this empirical test did not support mutation accumulation as a genetic mechanism contributing to the evolution of late-life mortality rate plateaus, it was not necessarily refuted, because different patterns of dominance among alleles with effects specific to late life could eliminate hybrid vigor (cf. Charlesworth and Hughes 1996). In any case, the absence of hybrid vigor is interesting in itself, because it suggests an absence of inbreeding depression in these populations.

While these five B populations experienced independent evolution for 18 years before this experiment was performed, their demographic selection regimes were identical. Our expectation was that, while the B populations evolved under the aegis of the same demographic selection, mutation accumulation might have produced enough divergence among the five populations to give hybrid vigor upon crossing. However, this is not what we found.

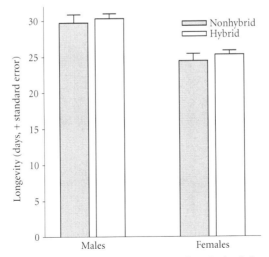

Figure 5-3. Mean longevity of male and female B flies derived from 14 hybrid and 5 nonhybrid crosses of the B populations.

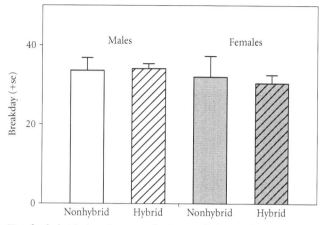

Figure 5-4. Test for hybrid vigor between the B populations. This figure shows the mean estimated mortality-rate plateau breakday for males and females from the nonhybrid (parental) and hybrid crosses, or the day on which a slope of zero better describes the mortality rate data than a nonzero slope. There was no significant difference between the breakdays of the hybridized and nonhybridized B cohorts (males: t-test, $p = 0.67$; females: t-test, $p = 0.46$). Error bars are standard errors.

RESULTS FROM TESTING WHETHER ANTAGONISTIC PLEIOTROPY ACTS AS A GENETIC MECHANISM IN THE EVOLUTION OF LATE-LIFE MORTALITY BY REVERSE EVOLUTION

To test antagonistic pleiotropy as a genetic mechanism involved in late-life mortality, we derived the five NRO populations from the corresponding O populations by reverse evolution and cultured them with an age of reproduction of 14 days from egg. The NRO populations were subjected to selection for early reproduction for only 24 generations after derivation from the O populations (see Figure 5-1) prior to their employment in the experimental assays described here. As we explained before, 24 generations is not enough time for either mutation accumulation or genetic drift to have a significant effect within populations of this size. After the 24 generations of selection for an earlier last age of reproduction, the five NRO populations were then compared to the five corresponding O populations from which they were derived with regard to the onset of late life. Evolutionary theory predicts that the NRO populations would eventually evolve an earlier plateau in mortality rates when compared to the Os. However, this pattern would be observable after so few generations of reverse selection only if antagonistic pleiotropy is operating as a genetic mechanism underlying the evolution of late life.

When the NRO populations were compared to their corresponding O populations using the same mortality assays that we previously described (Chapter 4

Table 5-1. Results from a test for response to a brief period of reverse selection in the NRO populations. Because each NRO population derives from a single O population, paired-difference *t*-tests were used to test for significant average differences between the O and NRO groups.

	Males		Females	
	O	NRO	O	NRO
Sample size	7,343	9,072	12,784	13,445
Breakday	68.6	48.2**	67.8	54.6[†]
Plateau mortality rate	0.28	0.22	0.24	0.26
A	0.0015	0.0021	0.0019	0.0033*
α	0.062	0.081*	0.063	0.076
Longevity	53.3	41.8**	50.4	39.2**

** $p < 0.01$; * $p < 0.05$; [†] $p < 0.1$.

and Rose et al. 2002), we found significant evidence of a rapid response to selection in the NRO populations with regard to the onset of mortality-rate plateaus, or the breakday (Table 5-1). In fact, the breakdays' response to selection was remarkably rapid and highly significant in males, demonstrating a net response of more than 20 days in only 24 generations. Furthermore, the female response to selection for the start of the mortality-rate plateau was nearly significant and showed a net response of 13 days in the predicted direction. Together these results are consistent with an evolutionary model in which the last age of reproduction and the evolution of mortality-rate plateaus are positively related (Figure 5-1). The highly significant male result with respect to the breakday after such a small number of generations of reverse selection is sufficient to support antagonistic pleiotropy as a genetic mechanism involved in the evolution of late-life mortality, as drift is unlikely to contribute a significant effect in populations of this size in such a short amount of evolutionary time.

DETERMINING WHETHER ANTAGONISTIC PLEIOTROPY ACTS AS A GENETIC MECHANISM IN THE EVOLUTION OF LATE-LIFE FECUNDITY

Because the start of late-life fecundity plateaus depends on the timing in the drop in fecundity's force of natural selection, specifically the last age of survival in the population's evolutionary history, we predicted that switching to a selection regime with an earlier last age of reproduction should lead to an earlier age for the onset of fecundity plateaus if antagonistic pleiotropy is a genetic mechanism underlying late-life fecundity patterns. This experimental design is analogous to the experiment that we performed to test whether antagonistic pleiotropy operates as a genetic mechanism in the evolution of late-life mortality rate plateaus. As with the mortality experiment, we subjected later-reproducing populations, specifically the CO populations described in Chapter 4 and above (see Figure 5-1), to an evolutionary reversion to earlier ages of reproduction (cf. Rose et al. 2002, 2004), and consequently, earlier ages of last survival. These newly derived populations were named NRCO, and each of these populations was derived from the corresponding CO population.

After the new selection regime had been imposed on the NRCO populations for only 24 generations, we compared the fecundity patterns of each of the new earlier-reproducing populations to its parental later-reproducing CO population to see whether the NRCO populations also had an earlier age of fecundity plateau onset (see Rauser et al. 2006b for experimental details). This selection regime selected not only for early reproduction in the NRCO populations, but also for accelerated development and an earlier last age of survival. As before, antagonistic pleiotropy is distinguished from other population genetic effects in this assay by implementing this specific experimental design, which allows for too little evolutionary time for mutation accumulation or drift to have a significant effect within the population sizes we employ.

Evolutionary theory predicts that the NRCO populations will evolve an earlier age of onset for the plateau in fecundity, compared to the CO populations, if antagonistic pleiotropy is a genetic mechanism shaping late-life fecundity patterns. Because antagonistic pleiotropy does not simply depend on early fecundity, selection for multiple early fitness characters (e.g., early reproduction or accelerated development) in the NRCO populations encompasses all types of antagonistic pleiotropy. Therefore, a shift in late-age fecundity in response to this selection will implicate antagonistic pleiotropy as a genetic mechanism shaping late-life fecundity patterns, regardless of which particular early-life fitness components are involved. Our pairwise comparisons of the late-life fecundity plateau patterns between the NRCO and CO populations corroborated this theory (statistical analyses were performed as described in the Appendix). Figure 5-5 depicts the average population fecundity for the four pairwise comparisons between these populations, along with the estimated ages of each breakday, or start of the late-life fecundity plateau.

The breakday, estimated from the two-stage model described in the Appendix section for Chapter 4, was significantly earlier in the earlier-reproducing NRCO populations compared to the later-reproducing CO populations (Table 5-2). This result suggests that late-life fecundity plateaus respond rapidly to selection

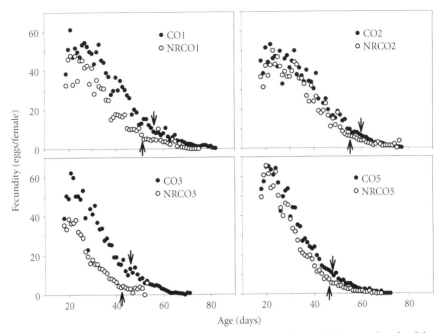

Figure 5-5. The mean age-specific fecundity for four NRCO populations and each of the CO populations from which they were derived. The arrows pointing down indicate the estimated breakday for the CO populations, while the arrows pointing up indicate the breakday for the NRCO populations. In each case the breakday occurs at a younger age in the earlier-reproducing NRCO population than in its paired later-reproducing CO population (mean breakday = 49.59 days in the NRCO populations and 56.16 days in the CO populations).

and that antagonistic pleiotropy connects late-life fecundity to early-life fitness characters, resulting in the evolution of an average pairwise difference of 6.57 days in only 24 generations.

Although our experimental results implicate antagonistic pleiotropy in the evolution of late life, it is important to note that the two genetic mechanisms of antagonistic pleiotropy and mutation accumulation are not mutually exclusive and that a positive result for antagonistic pleiotropy does not necessarily mean that mutation accumulation is not involved in the evolution of late life.

ANTAGONISTIC PLEIOTROPY IS IMPLICATED AS A GENETIC MECHANISM IN THE EVOLUTION OF LATE LIFE

With antagonistic pleiotropy between early and late ages, some of the alleles that enhance early reproduction will depress later survival or fecundity

Table 5-2. Results from the test of antagonistic pleiotropy demonstrating that the onset of the late-life fecundity plateau starts significantly earlier in the NRCO populations, selected for earlier reproduction for just 24 generations, compared to the CO populations. Plateau height was computed from Equation A4-2. The regression coefficients (see Equation A4-1) were estimated from the age-specific fecundity observed in 100 vials (400 females) randomly sampled daily from an initial population size of 3,200 vials (12,800 females).

	Population	
	NRCO	**CO**
Sample size	9,750	13,587
First-stage y-int (c_1)	71.24	84.80
First stage slope (c_2)	−1.36	−1.44
Breakday (*fbd*)	49.59	56.16***
Plateau height (φ_4) (eggs/female/day)	3.97	3.94

*** $p < 0.0001$.

(Williams 1957, 1966; Rose 1985; Charlesworth 1994). Furthermore, late-acting deleterious genes that cause reproductive senescence late in life can persist in a population because these same genes enhance reproduction, or other fitness characters, at earlier ages when the force of natural selection is much stronger.

We specifically tested antagonistic pleiotropy as a genetic mechanism affecting late-life mortality and fecundity in two independent reverse selection experiments. These experiments utilized our well-established later-reproducing O and CO populations (Figure 5-1). From these populations we used reverse selection for only a small number of generations to create new earlier-reproducing populations, which we then compared to their corresponding parental population. Evolutionary theory predicts that reversion to an earlier age of reproduction shifts the age at which the force of natural selection acting on fecundity and survival declines to zero. Natural selection on early reproduction for a short period will therefore tend to increase mortality rates and decrease fecundity later in life, provided that there is antagonistic pleiotropy between early and late ages.

The experiments on the genetic mechanisms of aging and late life that we have described in this chapter reveal that both late-life mortality and late-life fecundity can be remarkably responsive to selection for early reproduction imposed for a small number of generations, which implicates antagonistic pleiotropy as a powerful genetic mechanism shaping the evolution of late life. The forgoing studies demonstrate that antagonistic pleiotropy between early fitness-related characters and late-life characters can affect the evolution of late life. While no evidence for the action of mutation accumulation was found, there is no critical evidence against its involvement in the evolution of late life either.

Demography of Late Life with Lifelong Heterogeneity

Within-cohort selection can theoretically lead to the deceleration of mortality rates when there is substantial lifelong heterogeneity in robustness. This effect arises in both nonaging and aging organisms. If there is implausibly extreme lifelong heterogeneity in robustness, late life can arise from the relictual survival of the extremely robust.

THE CONCEPT OF LIFELONG DEMOGRAPHIC HETEROGENEITY

The first theories proposed to explain the leveling of mortality rates at late ages were not evolutionarily based, but instead were demographic theories based on lifelong differences in individual robustness within an aging cohort. These theories suppose that there is sufficient heterogeneity in lifelong robustness within a population to cause the slowing of mortality rates at late ages. That is, it is imagined that mortality rates will start to slow at later ages, after the less robust individuals in the population have died. Note that demographic heterogeneity should not be confused with mere genetic or environmental variation within a population (cf. Carnes and Olshansky 2001). The assumption of consistent life-long differences between individuals is more exigent than that.

The idea of demographic heterogeneity predates the definitive demonstration of late-life mortality-rate plateaus by Carey et al. (1992) and Curtsinger et al. (1992). In crude verbal form, the idea is mentioned by Greenwood and Irwin (1939). Beard (1959) derived mathematical models for mortality that accounted for lifelong heterogeneity in individual mortality. He was an actuary who primarily analyzed human data and was concerned about the way late-age human data did not conform to the Gompertz family of mortality models. Specifically, his mortality models included variables that incorporated individual differences in "vitality" (Beard 1964). He even suggested that these differences in individual vitality may be the underlying cause of the slowing in late-age human mortality rates within a population (Beard 1971).

However, it wasn't until Vaupel et al.'s (1979) publication that the first complete *lifelong demographic heterogeneity* theory to explain late life was developed, also based on observations made on human mortality data. This theory leads to a robust prediction of decelerating age-specific cohort mortality late in life, granting only a few, seemingly natural, assumptions. The Vaupel heterogeneity theory assumes that aging cohorts are comprised of a collection of secondary groups, with each subgroup having its own characteristic Gompertz function that defines its mortality pattern. Thus, one subgroup might have a relatively low baseline mortality rate (*A* from Equation 2-1) compared to other subgroups that will reduce its age-specific mortality rates throughout life, but the same rate of aging (*α* from Equation 2-1). With this version of Vaupel's heterogeneity model, the average age-specific mortality rate is

$$\bar{\mu}(x) = \frac{Ae^{\alpha x}}{1 + \left[\sigma^2 A\left(e^{\alpha x} - 1\right)\right]\alpha^{-1}}, \tag{6-1}$$

where σ^2 is proportional to the variance in *A*. At advanced ages, once most individuals in the less robust subgroups have died, the average mortality rate of Equation 6-1 approaches a mortality-rate plateau equal to $\alpha\sigma^{-2}$.

On the other hand, the rate of aging, or the value of *α*, may be the parameter that is imagined to vary among the subgroups, resulting in some groups having a significantly higher rate of aging than others (e.g., Pletcher and Curtsinger 2000). Although allowing the rate of aging to vary among such hypothetical subgroups is much more difficult to analyze, Pletcher and Curtsinger (2000) and Service (2000) have examined the age-dependent changes in the variance of mortality rates with models of this kind. Note that with this general type of model, regardless of whether the variation in mortality lies within the baseline mortality rate (*A*) or the rate of aging (*α*), the hypothetical differences among the subgroups are lifelong. That is, individuals that are less robust at late ages are imagined to be less robust at all other ages, too.

SOURCES OF VARIATION

Sources of possible variation in mortality are outlined in Figure 6-1.

At the first level, genetic variation may affect the age-independent and age-dependent parameters of the Gompertz equation. We examine the ith genotype in Figure 6-1 in more detail. Individuals that are identical for this genotype may also

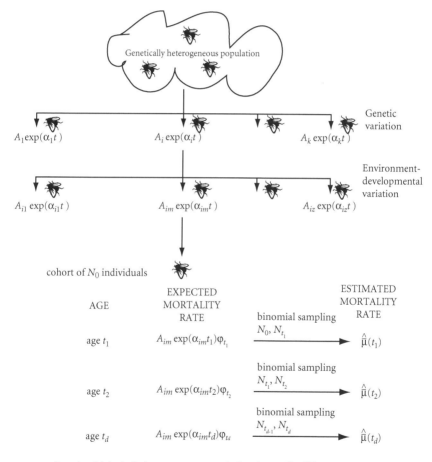

Even in a biologically homogeneous population there will still be variation in the estimated mortality rates due to binomial sampling and experimental error (φ_{t_i}).

Figure 6-1. Potential sources of variation in mortality rates estimated from experimental cohorts include genetic variation (subscript i) and environment-developmental variation (subscript m).

vary in their mortality rates due to environmental differences encountered early in life, which may cause A, α, or both variables to vary over the individuals' lifetime. Environment, m, in Figure 6-1, for example, can be examined in more detail. Suppose we could create many individuals of genotype i that have all experienced the exact same environment, m. Although this would be difficult in practice, it could be accomplished in computer simulations. However, even under these conditions, theoretically there is expected to be variation in the estimated mortality rates due to binomial sampling variance and experimental error. That is, for genotype i in environment m the chance of an individual surviving to age t_1 is

$$p(t_1) = \exp\left\{\frac{A_{im}\varphi_{t_1}\left[1 - \exp\left(\alpha_{im}t_1\right)\right]}{\alpha_{im}}\right\}$$

Thus, the expected number of survivors at age t_1 is $N_{t1} = p(t_1)N_0$, which has a binomial distribution with a variance equal to $p(t_1)[1-p(t_1)]N_0$.

EFFECTS OF EXPERIMENTAL ERROR

We can explore the extent to which experimental error may be responsible for mortality rate plateaus by generating artificial cohorts of fruit flies on the computer with varying levels of experimental error.

In Figure 6-2, mortality is simulated in cohorts of 1,000 individuals with Gompertz parameters $A = 0.00725346$ and $\alpha = 0.22891005$ (see the Appendix for details). These values were estimated from the mortality of a large number of *Drosophila* during ages prior to the mortality plateau. The median longevity of the simulated populations in Figure 6-2 is just 14 days, and deaths are estimated to the nearest day. The experimental error is assumed to have a normal distribution with a mean of zero and standard deviations ranging from zero to six, as given in Figure 6-2. For small to moderate variance in experimental errors, there is no suggestion that the Gompertz mortality trajectory slows at later ages. However, there is a slight suggestion of slowing when the standard deviation reaches six. But this level of experimental error would mean that the estimated age at death would be mistaken by nearly 12 days, which is almost equal to the median longevity of individuals in these cohorts. These errors must arise from an extreme propensity to make mistakes, such as calling a fly dead when it is in fact alive or handling a fly in such a way as to cause its premature death. Yet it is highly unlikely that, in experienced hands, experimental error would ever be this large. In conclusion, variation caused solely by experimental error is highly unlikely to contribute substantially to the mortality rate plateaus

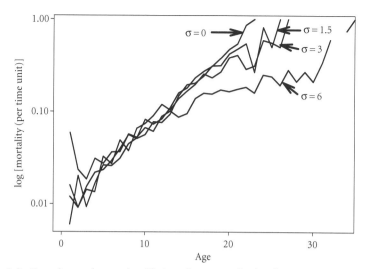

Figure 6-2. Experimental error is added to the age at death of an experimental cohort. These errors are assumed to have a normal distribution with mean zero and variance σ^2. Thus, when $\sigma = 6$, the 95% confidence interval on the estimated age at death is ±12 days.

that have been observed in a variety of organisms, especially under controlled laboratory conditions.

SIMULATED EFFECTS OF HYPOTHETICAL EXTREME LIFELONG HETEROGENEITY FOR *A* AND α

Empirically, there are several environmental factors that are known to affect longevity. One is temperature; however, this is unlikely to be important in laboratory populations, where this variable can be carefully controlled. The other factor is food level or caloric restriction, and it is certainly possible that individuals might vary in their food intake in laboratory experiments. Furthermore, in *Drosophila* there is good evidence that caloric restriction increases longevity through a decline in the age-independent parameter of the Gompertz equation, *A* (e.g., Nusbaum et al. 1996). On a natural log scale, caloric restriction results in a decline in *A* of about 23%. It seems unlikely that, in a carefully controlled environment, subtle differences between the environments would be sustained long enough to result in changes in *A* much larger than those that are purposefully induced. In any case, we can examine whether late-life mortality-rate plateaus can be generated by producing such extreme variation in *A*.

In Figure 6-3a, we show the natural log of mortality for a *Drosophila* B-female cohort (see Figure 4-2 for population description) along with mortality rates from several different computer simulations. Even when the variation in *A* is as large as the variation produced by deliberate and extreme caloric restriction in *Drosophila* experiments (the aforementioned 23%), the mortality rates are almost indistinguishable from the standard Gompertz model with 0% lifelong heterogeneity.

When we make the variation in *A* four times greater than the variation associated with lifelong caloric restriction, a slight plateau is visible, although it is still not as pronounced as the plateau actually observed in B-female cohorts. Note that this is a hypothetical environmental effect that is vastly greater than any yet detected in a *Drosophila* experiment. Nonetheless, this effect still isn't big enough to generate the late-life mortality-rate plateaus that have actually been observed in *Drosophila* cohorts.

Hypothetical lifelong heterogeneity models may also include variation in the age-dependent parameter of the Gompertz equation (α). In *Drosophila*, environmental effects don't typically affect the α parameter, but genetic changes may (Nusbaum et al. 1996). Long-term natural laboratory selection over many generations has resulted in a 19% (on a natural log scale) reduction in the α parameter within the long-lived O populations relative to their controls, the Bs (see Figure 4-2). Figure 6-3b simply repeats the hypothetical calculations that were done in Figure 6-3a, but with hypothetically extreme variation in α rather than *A*. Even theoretically generated cohorts with lifelong heterogeneity as large as the difference between the B and O populations, which produces two- to threefold differentiation in average longevity, do not exhibit late-life mortality-rate plateaus. The dashed line in Figure 6-3b has artificially generated lifelong heterogeneity in both *A* and α parameters at magnitudes equal to those produced by caloric restriction (23%) and long-sustained natural selection level (19%), respectively, giving an effect that is not qualitatively discernible in the plot from the cases with either one of these individual assumptions.

In this example, it is important to bear in mind that actual B-female cohorts have extremely few individuals that live as long as the top 30% longevities from O populations, with or without caloric restriction. So, this unreasonably favorable scenario for the lifelong heterogeneity hypothesis is clearly erroneous even as a bare proposition; it does not correspond to what is ever observed.

But even with such extreme hypothetical variation in both *A* and α parameters, age-specific mortality rates in simulated cohorts fail to plateau at late ages. With sufficiently large synthetic variation in α (100%), plateaus do eventually appear in our entirely hypothetical simulated cohorts. However, no conditions, whether genetic or environmental, have yet been identified empirically that could reasonably be expected to yield variation in α of this extreme magnitude,

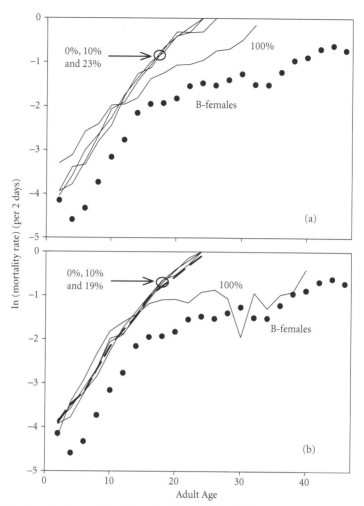

Figure 6-3. Simulated mortality with entirely hypothetical lifelong individual heterogeneity for the A and α parameters of the Gompertz equation (Equation 2-1). In panel (a) of the figure, the natural log of A was assumed to have a log normal distribution. Observed mortality data for actual B-female cohorts are shown as circles. The lifelong heterogeneity variances were chosen so that a 95% confidence interval was equal to 0, 10, 23, or 100% of the mean value of A. Simulated cohorts were the same size as the actual B-female cohorts. The values of A and α used in the simulations were estimated from the first 24 days of the actual B-female data. During the first 24 days these cohorts age according to the Gompertz equation, and after this age they start to plateau (data from Rose et al. 2002). Simulation results for panel (b) of the figure were carried out as described for panel (a), except that the artificially imposed lifelong heterogeneity was in the age-dependent parameter of the Gompertz (α). The solid lines represent the simulations with 0, 10, 19, and 100% variation in α, while the dashed line is with 23% variation in A and 19% variation in α. Details of these simulations are given in the Appendix section for this chapter.

even when experimental evolution over hundreds of generations is deliberately used to force the greatest possible differentiation in these parameters.

LIFELONG HETEROGENEITY FOR FECUNDITY

There is no equally natural explanation of late-life plateaus in fecundity that derives directly from extant lifelong heterogeneity theory, but post hoc explanations are always possible with a theory as ill-defined and open-ended as lifelong heterogeneity theory. One such explanation could be the differential loss of more fecund individuals. That is, it is conceivable that some females lay a lot of eggs at early ages but die prematurely, leaving only those females that always laid a low number of eggs preponderant among the females still alive at later ages. This hypothetical scenario assumes that there is a trade-off between mortality and reproduction, and it couples high mortality with high fecundity, and conversely. Another possible heterogeneity explanation for the existence of late-life fecundity plateaus could be based on some sort of highly generalized robustness, whereby some females both survive better and are more fecund. In addition to these two possible explanations, any number of variations based on heterogeneity in fecundity can be imagined, given the wide latitude with which lifelong heterogeneity scenarios can be constructed.

SIMULATION OF LIFELONG HETEROGENEITY EFFECTS ON COHORT COMPOSITION FOR FECUNDITY

We have examined the consequences for average population fecundity of a cohort with two levels of robustness in fecundity and mortality (see Rauser et al. 2005a). We assumed that a phenotype with high fecundity was coupled with high mortality (H:H) and a phenotype with low fecundity was coupled with low mortality (L:L). A population consisting of just these two phenotypes is the simplest example of the trade-off version of a lifelong heterogeneity theory for fecundity. Specifically, we assume that more fecund individuals die earlier, leaving the less fecund individuals at later ages. We do not offer this example because we think that it is the only possible example of a theory of this kind. We are merely illustrating what the features of such theories are when they are formally explicit, in one case. Many models of this type can be invented in the wide-open context of lifelong heterogeneity theory.

We assumed that the H:H phenotype initially occurs at a frequency p, and thus L:L females are at a frequency of $1 - p$. We modeled adult survival with the Gompertz equation. The probability of survival to age t, l_t, is

$$l_t = exp\left\{\frac{A(1-exp(\alpha t))}{\alpha}\right\},$$

where A is the age-independent mortality parameter and α is the age-dependent mortality parameter. If we let the age-specific survival and fecundity of H:H females be l_t and m_t, respectively, and for L:L females $\tilde{l}_t$ and $\tilde{m}_t$, then the average fecundity of a cohort aged t days is

$$\frac{pl_t m_t + (1-p)\tilde{l}_t\tilde{m}_t}{pl_t + (1-p)\tilde{l}_t}. \tag{6-2}$$

The average population fecundity of a cohort with two levels of lifelong fecundity and mortality is high at early adult ages and decreases with age until it plateaus at low fecundity levels (Figure 6-4). This plateau in fecundity at late ages occurs once almost all of the lifelong, high-fecundity, high-mortality individuals have died. The results of this simulation demonstrate how such a hypothetical model with two levels of heterogeneity within a cohort can result in the average population fecundity patterns we have observed (Rauser et al. 2003, 2005a, 2005b, 2006b). However, we are not asserting that this is the only conceivable lifelong heterogeneity model that has such properties. In Chapter 8 we will review empirical tests of such fecundity heterogeneity models.

CONCLUSION: THE BIG STRAIN OF LIFELONG HETEROGENEITY THEORIES

The lifelong heterogeneity theories we have reviewed in this chapter do not rest upon well-established principles of biology and require extremely high levels of lifelong heterogeneity. The lack of a well-defined mechanistic basis for these heterogeneity theories makes it difficult to measure or infer the types of lifelong heterogeneity that these theories require. The chief support for these theories comes from their ability to mimic post hoc patterns of mortality seen in actual biological populations. This is a weak form of support for models in biology, because there are often many conceivable post hoc models with these properties (vid. Mueller and Joshi 2000, chapter 1). In this chapter, we have focused on demonstrating that extremely high levels of lifelong heterogeneity are required even to construct the type of hypothetical post hoc model that has been used to fit observed cohort survival patterns. In Chapter 7 we discuss whether such extreme levels of lifelong heterogeneity could plausibly evolve,

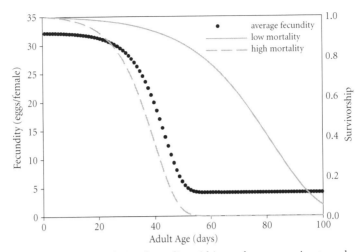

Figure 6-4. The average population fecundity within a cohort, assuming two phenotypes: high mortality with high fecundity (H:H, dashed gray line) and low mortality with low fecundity (L:L, solid gray line) using Equation 6-2. Average fecundity starts high and then declines with age until it stops declining at low levels at late ages. The onset of the plateau in average fecundity occurs once almost all H:H individuals have died. These results assume that the H:H and L:L types start at equal frequencies, $p = 0.5$. The A and α parameters were assumed to be 9.13×10^{-4} and 0.123, respectively, for the H:H females and 4.75×10^{-4} and 0.059, respectively, for the L:L females. These estimates were taken from actual mortality data from long- and short-lived fly populations (Nusbaum et al. 1996, Table 1). We assumed that H:H females had a constant high fecundity such that $m_t = 60$ eggs/day for all t and L:L females had a constant low fecundity with $\bar{m}_t$ eggs/day. The solid and dashed gray lines represent the proportion of individuals alive at each age (survivorship) for the high and low-mortality phenotypes, respectively.

and in Chapter 8 we discuss critical tests of lifelong heterogeneity theories. However, we hope that this chapter has already shown the reader the extent to which these theories impose considerable strain on biological credulity with respect to the magnitude of their presumed lifelong heterogeneity.

Many heterogeneity theories proposed to explain the slowing of mortality rates at late ages assume that individuals within a cohort are still aging according to Gompertz's law but that the differences between individual Gompertz functions is large (Vaupel 1990; Kowald and Kirkwood 1993). Like us, Abrams and Ludwig (1995) point out that the amount of heterogeneity assumed to make these models fit population mortality-rate data is extremely large, without precedent in actual data. In fact, the difference between the Gompertz parameters in our long- and short-lived fly populations (Nusbaum et al. 1996) does not come close to the magnitude of heterogeneity required within a population to make heterogeneity models fit the data (Kowald and Kirkwood 1993; Vaupel and Carey 1993). Furthermore, we know that the demographic patterns of short-lived populations do not indicate the presence of individuals as long-lived as the typical member of the long-lived populations that we have studied.

Evolution of Lifelong Heterogeneity

It is unlikely that lifelong heterogeneity in robustness will be extreme, if it occurs at all. Natural selection will favor genotypes with much greater lifelong robustness, reducing the genetic variance for robustness over time. Natural selection will also favor genotypes that reduce the amount of developmental or environmental variation in robustness. Stable genetic equilibria with sufficient levels of lifelong heterogeneity to cause mortality plateaus seem unlikely.

AN EVOLUTIONARY CRITIQUE OF LIFELONG HETEROGENEITY THEORY

Some variability in robustness, the underlying controller of mortality rates in cohorts free of exogenous mortality, undoubtedly exists within natural populations due to genetic and environmental variation. In fact, there is a substantial literature showing that life-history characters vary (reviewed in Finch 1990; Rose 1991; Roff 1992; Stearns 1992), which might be taken to mean that the lifelong heterogeneity model is well founded. Indeed, some types of heterogeneity can arise when evolution by natural selection maintains genetic variation. But what the heterogeneity theory requires, as an explanation of the profound late-life deceleration of aging, is sufficiently extreme lifelong differences in individual mortality rates to produce this effect. This is the essential problem: for lifelong heterogeneity to work as an explanatory hypothesis for late-life phenomena, it must be so extreme that it raises the question of whether or not it is even remotely plausible.

The lifelong heterogeneity theory of late life faces major difficulties in meeting this challenge, not only with respect to observable data, but even with respect to basic theoretical presuppositions. Evolutionary theory predicts that natural selection will tend to decrease genetic variation in fitness-related traits like early adult mortality (Nagylaki 1992), because genetic variation in fitness is the "fuel" that natural selection consumes to produce adaptation. Yet the lifelong heterogeneity theory requires a large amount of sustained lifelong heterogeneity for mortality, either genetic or environmental. If the genetic heterogeneity for mortality rates, both early and late, is heritable, it will be strongly subject to natural selection. Natural selection will, of course, reduce the amount of genetic heterogeneity in a population over time, unless there is some form of balancing selection, which is not necessarily common. Without balancing selection, or some other mechanism constantly introducing genetic variation for fitness into the population, natural selection will purge most genetic variation for lifelong robustness from the population.

The lifelong heterogeneity required by the late-life heterogeneity theory can also be environmental or even merely developmental. If there is a substantial amount of variation arising from the environment, whether it is spatial, temporal, or both, then the measure of fitness is given by the average effect of an allele minus a term giving the variation in fitness. Thus, the equation $\mu - \frac{1}{2}\sigma^2$ determines the evolutionary outcome, or fitness, of a genotype (Gillespie 1973), where μ is the measure of average fitness and σ^2 is the measure of environmental variance in fitness. Therefore, evolution by natural selection will also tend to reduce environmental sources of lifelong heterogeneity. Because fecundity is also a major fitness component, we expect the same reduction in genetic and environmental variation from natural selection on fecundity as with mortality.

SIMULATED EVOLUTION OF LIFELONG HETEROGENEITY

To illustrate the potential of evolution to maintain genetic variation, and to evaluate the likelihood that such genetic variation will lead to substantial lifelong heterogeneity, we studied a simple single-locus population genetic model. We started the population at a state of complete fixation for a single allele that determined a particular Gompertz mortality phenotype in the

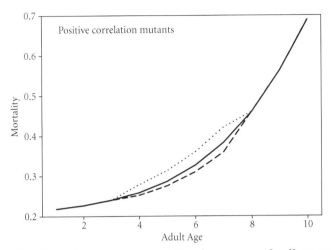

Figure 7-1. The effects of mutation on mortality when age-specific effects are positively correlated, in keeping with lifelong heterogeneity theory. The solid line is the initial starting Gompertz mortality. The dotted line shows a mutant phenotype with deleterious effects. The dashed line shows a mutant phenotype with beneficial effects. See the Appendix section for this chapter for more details.

homozygous carriers of this allele (see Figure 7-1). Mutations were then generated that affected the mortality phenotype of the most common allele in the population in two different ways.

In the first scenario that we considered, mutants were constrained to have positive correlations for age-specific effects (Figure 7-1), and these effects could be either entirely positive or entirely negative. This is the case assumed by genetically based models of lifelong heterogeneity.

Alternatively, in our second scenario, mutations could have a negative correlation with respect to their effects on age-specific mortalities. Thus, a mutant that had an increase in mortality at an early age would have the pleiotropic effect of decreasing mortality at later ages (Figure 7-2). Under these conditions, there is no lifelong heterogeneity, though there can be age-specific variation.

For each of the mutation schemes outlined in Figures 7-1 and 7-2, we generated 100 mutants. After the creation of each new mutant and its associated fitness, we used standard single-locus population genetic theory to determine the outcome of evolution (see the Appendix section for this chapter for further details). Allele frequencies were iterated for 50,000 generations or until an equilibrium was reached, whichever came first. The mean fitness and number of alleles maintained by selection are shown in Figures 7-3 and 7-4.

Although the simulations with positive correlations among genetic effects on survival across ages produced two and occasionally three allele polymorphisms,

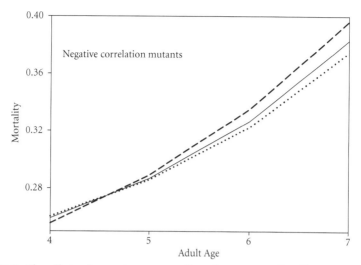

Figure 7-2. The effects of mutation on mortality when age-specific effects are negatively correlated. The solid line is the initial starting Gompertz mortality. The dotted line shows a mutant phenotype with deleterious effects early in life and beneficial effects later. The dashed line shows a mutant phenotype with beneficial effects early in life and deleterious effects later. See the Appendix for more details.

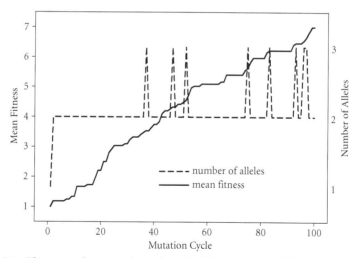

Figure 7-3. The mean fitness and number of alleles maintained by selection after the introduction of 100 mutants into a population initially showing Gompertz mortality patterns. These mutants showed positive correlations in age-specific mortality changes (see Figure 7-1). Between the introductions of each new mutant, selection was allowed to progress for up to 50,000 generations (see the Appendix for more details).

the two allele polymorphisms typically had a single very common allele, and the rare allele declined in frequency over the course of selection after it arose (Figure 7-3). With negative correlations, there were typically six alleles maintained by selection (Figure 7-4). This particular simulation ended with a seven-allele polymorphism. This was not a transient state, since we checked the equilibrium allele frequencies and stability conditions by standard analytical procedures (Mandel 1959); this seven-allele equilibrium was globally stable, and the computer simulation had accurately converged to the equilibrium allele frequencies. [Previous research on randomly generated fitness matrices has shown that selection at a single locus can typically support six- and seven-allele polymorphisms (Spencer and Marks 1992), so the observations in Figure 7-4 are not exceptional in that regard.] In other words, when there is the type of positive correlation in genetic effects required to generate lifelong heterogeneity, there is systematically less genetic variation than there is when there are negative correlations among genetic effects on survival.

In Chapter 6, we showed that the levels of lifelong heterogeneity in demographic parameters must be exceptionally large to generate late-life plateaus. The present simulation results provide another means of testing whether or not this requirement is likely to be met. If we use the final equilibrium allele frequencies and phenotypes from the evolutionary process shown in Figure 7-4,

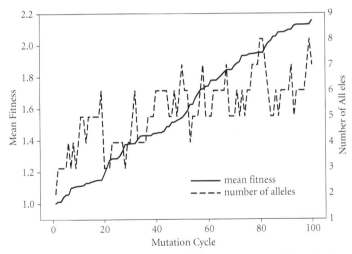

Figure 7-4. The mean fitness and number of alleles maintained by selection after the introduction of 100 mutants into a population initially showing Gompertz mortality patterns. These mutants showed negative correlations in age-specific mortality changes (see Figure 7-2). Between the introductions of each new mutant, selection was allowed to progress for up to 50,000 generations (see the Appendix for more details).

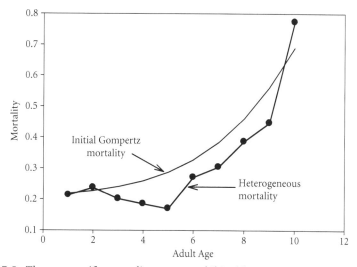

Figure 7-5. The age-specific mortality pattern exhibited by a genetically heterogeneous cohort that has evolved by modifying an initial Gompertz mortality pattern. The allele frequencies and genotypic specific mortalities were derived from the final equilibrium population shown in Figure 7-4. The age at death of 1,566 individuals at Hardy-Weinberg equilibrium were simulated to give the mortality values in the "Heterogeneous mortality" curve above. Since this population had seven alleles, there were a total of 28 genotypes that were represented with at least one individual in this simulation.

we can simulate mortality in this heterogeneous population and see if the late-life mortality level off. The results from such an exercise are shown in Figure 7-5. Note that this model is limited to genetic effects at one locus on an underlying Gompertzian mortality pattern. Its failure to generate late-life plateaus is not evidence against Hamiltonian theories of late life, which are not subject to this type of evolutionary constraint.

Despite the genetic heterogeneity due to the seven-allele polymorphism, there is no leveling of mortality in late life. This result is in part a consequence of the magnitude of the genetic differences between the different mutants. However, as new mutants with ever-decreasing mortality are introduced into the simulated population, alleles with substantially higher mortality are gradually eliminated from the population despite the overdominance built into this genetic model (see the Appendix for details). This scenario suggests that the buildup of the large-scale genetic variation required to create late-life mortality plateaus is unlikely to occur in populations that evolve according to the rules of population genetics. This does not, of course, preclude the artificial generation of lifelong heterogeneity in composite cohorts assembled from extremely different genotypes, as first supposed by Greenwood and Irwin (1939). An omnipotent deity could do this for a population in the wild or an interfering gerontologist could do this for a model organism in the laboratory. But calculations like these suggest that such lifelong heterogeneity is not readily or ubiquitously produced by natural processes occurring on their own.

CONCLUSION: EVOLUTION ABHORS EXTREME LIFELONG HETEROGENEITY

Demographic heterogeneity may arise from standing genetic variation in natural populations. However, these explanations must confront the problem that, when such genetic variation has lifelong, pronounced, and positively correlated effects on components of fitness like mortality and fecundity, it is certain to affect fitness and thus be acted upon by natural selection. Population genetic theory suggests that, all other things being equal, natural selection will favor reductions in fitness variation, making the lifelong heterogeneity hypothesis less tenable as an explanation for the pronounced deceleration of aging that is observed during late life. Although certain patterns of genetic variation, like overdominance with antagonistic effects across ages—the opposite pattern from lifelong heterogeneity—may result in stable genetic polymorphisms, our results suggest that such population-genetic mechanisms are unlikely to sustain the type of variation that could produce late-life mortality plateaus from the sieving of lifelong heterogeneity.

Experimental Tests of Lifelong Heterogeneity

Experimental tests of lifelong heterogeneity generally do not corroborate the theory. For example, populations that have evolved a considerable increase in the level of their robustness show little change in their late-life patterns compared to their parental populations, and physiological manipulation of cohort heterogeneity does not significantly affect the occurrence of late life. Furthermore, lifelong heterogeneity models predict that far more individuals survive to very late ages than are observed when these models are fit to actual data. With fecundity, experimental studies have shown that heterogeneity is not lifelong; that is, heterogeneity in early fecundity does not predict late-life outcomes.

TESTS OF LIFELONG HETEROGENEITY IN MORTALITY BASED ON REDUCED VARIANCE

Recall from Chapter 6 that the Vaupelian heterogeneity theory based on lifelong differences in robustness requires a large amount of variance in A or α

values between subgroups of individuals comprising a cohort. Although life-long heterogeneity this extreme has yet to be shown experimentally for any organism, a theoretical analysis of the Carey et al. (1992) mortality data for medflies demonstrated that their data could be fitted post hoc to a Vaupelian demographic heterogeneity model (Kowald and Kirkwood 1993), using entirely hypothetical high levels of lifelong heterogeneity.

So, where might this extreme lifelong heterogeneity in mortality come from? In general, it could be genetic or environmental in origin. Therefore, if the hypothesized lifelong heterogeneity is genetic, then an important empirical cor-ollary is that genetically homogeneous populations should show a less distinct mortality-rate plateau compared to genetically heterogeneous populations. To this end, Brooks et al. (1994) compared an isogenic cohort of *Caenorhabditis elegans* with a cohort that they deliberately constructed from extremely differ-ent mutants. Naturally enough, they found a more distinct plateau in the het-erogeneous population, just as Greenwood and Irwin (1939) had suggested would be found in such contrived cohorts 55 years earlier, based on the demog-raphy of *Drosophila* mutants. But Vaupel et al. (1994) pointed out that the iso-genic line was grown under different environmental conditions than the heterogeneous line, complicating the interpretation of these results.

It is possible that genetic variation could drive the heterogeneity models. But extensive experimental work has shown that, after removing genetic variation by extensive inbreeding, well-defined late-life mortality-rate plateaus continue to be observed (Curtsinger et al. 1992; Fukui et al. 1993, 1996). In particular, Fukui et al. (1993) found clear mortality plateaus with highly inbred *Drosophila* lines (inbreeding coefficient >0.99), suggesting that genetic variation is not required for mortality plateaus to occur. In any case, as we demonstrated in Chapter 7, ge-netic variation is not a plausible source of the hypothetical, extreme, and lifelong heterogeneity in mortality required to explain the existence of late-life plateaus in mortality, since evolution would strongly favor the elimination of genetic variants associated with such extreme and consistent differences in death rates. It follows then that, in the absence of genetic variation, all lifelong heterogeneity that is sup-posed to cause late-life mortality-rate plateaus must be environmental in origin.

Thus, if lifelong heterogeneity in mortality does not arise from genetic hetero-geneity, then it must come from heterogeneity in the environment or from acci-dents of development. However, Khazaeli et al. (1998) found that heterogeneity that was environmentally induced in flies is not a primary factor in determining late-life mortality rates. They went to a great deal of trouble to reduce recondite sources of variation in laboratory-reared cohorts of inbred *Drosophila melano-gaster* lines, and compared cohorts handled so as to reduce environmental varia-tion with cohorts in which no such care was exerted. Through diligent application of this procedure in the experimental cohorts, they were able to reduce the variance in age at death in these cohorts compared to unmanipulated controls, indicating that they had successfully reduced environmental sources of variation. This was achieved with two different inbred lines, as well as two different conditions of cohort maintenance, with and without mates. While there were somewhat fewer experimental, or low-variance, cohorts (64 out of 69) which showed mortality-rate deceleration compared to the control, or high-variance, cohorts (37 out of 37), this effect was not statistically significant. In particular, strenuous attempts to reduce environmental variance during larval and pupal development did not come close to abolishing the transition to late life consistently, any more than severe inbreeding has (vid. Fukui et al. 1993). Apparently, these efforts did reduce the initial mortality level in virgin cohorts (Khazaeli et al. 1998), which may in turn have had some effect on the ability of the experimenters to detect the transition to late life in experimental low-variance cohorts. If the Gompertz demographic parameters are affected by environmental effects, then plateaus should have been less prominent or nonexistent in the reduced-variance populations. However, there was also no difference found in the timing of late-age mortality deceleration between these populations, further suggesting that variation in the preadult environment contributes little to the creation of lifelong heterogeneity in demographic parameters. They concluded that "environmental heterogeneity accrued during larval development is not a major factor contributing to mortality deceleration at older ages" (Khazaeli et al. 1998, p. 314). We know of no other experiments of this kind that have reached a different conclusion, regardless of the methods used to reduce environmental sources of variation.

TESTING LIFELONG HETEROGENEITY USING EXTREMAL SURVIVORS

Service (2000, 2004) demonstrated that the natural log of age-specific mortality rates should show a unimodal distribution if there is sufficiently large variation in A and α across genetically different populations to explain late life. We examined this variance across the five B populations and the five O populations (Mueller et al. 2003). Despite the fact that these populations had been isolated and undergoing independent evolution for 100–500 generations at the time of this experiment, the pattern predicted by Service was not seen. These observations don't preclude the possibility that purposeful methods of creating genetic differentiation between populations, like selection or inbreeding, might not result in these patterns. However, differentiation that arises naturally from random genetic drift is apparently not sufficient to cause these unimodal patterns.

Inspired by the analysis of Service (2000), Mueller et al. (2003) tested the Vaupelian heterogeneity theory by fitting lifelong heterogeneity models to mortality data from cohorts of D. *melanogaster*, specifically choosing parameter values for these models that fit the observed cohort data as closely as possible. One such model, the heterogeneity-in-α model, assumes that a small portion of the population will have very small values of α, or slow rates of aging, and will consequently be very long-lived. Service (2000) produced some calculations suggesting that this model is an adequate explanation of mortality plateaus in cohorts of *Drosophila*. When he varied α in his simulations, populations with average longevities of 50 days were generated, which is reasonable for D. *melanogaster*. But these simulations also resulted in maximum lifespans of 365 days in reasonably sized cohorts, which is absurdly long for this species. We know of no case of a D. *melanogaster* individual surviving as long as 200 days when adult diapause is not induced.

We explore these questions in more detail here and expand upon our earlier work from Mueller et al. (2003). In Chapter 6, we showed that if heterogeneity entered the Gompertz equation only through the age-independent parameter, then it is very difficult to generate sufficient variation to account for the plateaus in late-life mortality observed in *Drosophila* cohorts. This problem is somewhat reduced if heterogeneity is introduced in the age-dependent Gompertz parameter (α), so we consider this latter case in more detail here.

Consider a model in which the age-dependent parameter, α, is a random variable equal to $\xi\tilde{\alpha}$, where the random variable, ζ, has a gamma distribution with a mean of one and variance equal to k^{-1}. We call this the *heterogeneity-in-α model*. The mean (over all individuals with different α-values, i.e., α-types) instantaneous mortality rate for individuals aged x under the heterogeneity-in-α model is, following Pletcher and Curtsinger (2000), given by

$$\bar{u}(x) = \frac{\int_0^\infty A z^{k-1} exp\left[\tilde{\alpha}zx - \phi(x,z)\right] dz}{\int_0^\infty z^{k-1} exp\left[-\phi(x,z)\right] dz}, \tag{8-1}$$

where $\phi(x, z) = kz + A(\tilde{\alpha}z)^{-1}[\,exp(\tilde{\alpha}zx-1)]$. To estimate the three parameters in Equation 8-1, we need to fit the observed mortality over finite time periods of several days to the predictions of the model. The predicted mortality between times t_1 and t_2 $(t_2 > t_1)$ is given by $1 - \frac{P_{t_2}}{P_{t_1}}$, where p_t is the probability of surviving to time t. We estimated this heuristically as follows. If we let $\frac{-1}{N}\frac{dN}{dt} = \bar{u}(x)$, then it follows that

$$\frac{N_t}{N_0} = p_t = \exp\left\{\int_o^t \bar{u}(x)dx\right\} \tag{8-2}$$

Equation 8-2 can only be considered an approximation, because $\bar{u}(x)$ is an average mortality rate and thus the integral in Equation 8-2 is only an approximation to the average of the integrals of each of the different α-types in the population (e.g., the integral of the average mortality rate is not equal to the average of the integrals of the distinct mortality rates).

The weighted least-squares fit to the heterogeneity model for actual B1 female cohort data is shown in Figure 8-1a. (See the Appendix section for this chapter for the details of the estimation methods.) Given the variability of these observations, it would appear that the heterogeneity-in-α model can mimic the mortality rates of *Drosophila* quite well. However, the heterogeneity-in-α model is more than an equation that mimics these observed cohort deaths post hoc. It also contains a scientific hypothesis about the cause of these mortality rates. Therefore, it can be subject to more careful scrutiny than just goodness of fit.

The least squares estimates of the lifelong heterogeneity-in-α model parameters can be used to generate the distribution of age-at-death under this model (see the Appendix), which can then be compared to the observed distribution. Following this procedure for the B1 female cohort (Figure 8-1b), we see that a larger fraction of the cohort dies at younger ages than predicted by the heterogeneity-in-α model.

To formally test the ability of the heterogeneity-in-α model to predict the distribution of age at death in observed cohorts *when the model has been specifically fit to these particular cohorts*, we carried out two different statistical tests. We generated 100 sample populations using the parameter estimates from the heterogeneity-in-α model (see the Appendix section for Chapter 8 for the fits of this model to each of the 40 study populations). Each sample cohort was the same size as our original *Drosophila* population. We then used the Kolmogorov-Smirnov test to determine if the observed cumulative distribution function (CDF) was above the heterogeneity-in-α model CDF. This hypothesis test specifically addresses the previous observation that a larger fraction of the population dies at younger ages than predicted by the heterogeneity-in-α model.

For each of the 40 observed cohorts of *Drosophila*, in Table 8-1 we show the fraction of the 100 tests that resulted in a statistically significant difference between the observed and expected CDF from the heterogeneity-in-α model. We see that in the overwhelming majority of the populations, the heterogeneity-in-α model produces a significantly different CDF function, typically with more probability mass in the right tail of the distribution.

We have done a similar analysis focusing only on the tail of the distribution of age at death. We chose an age at which about 90% of the population is expected to be dead under the heterogeneity-in-α model. We then compared the observed frequency of the population still alive ($\bar{p}$) to the expected ($\hat{p}$) and used a binomial test to determine if $\bar{p} < \hat{p}$.

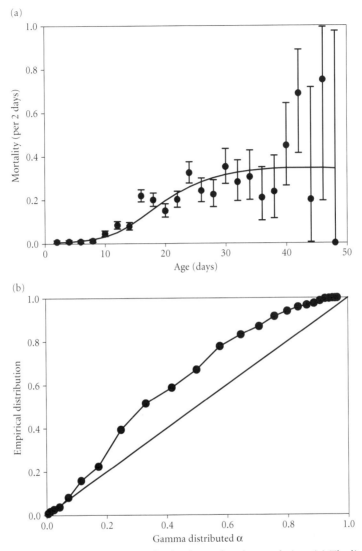

Figure 8-1. The two figures show results for the B1 female population. (a) The line is the weighted least-squares nonlinear fit of the heterogeneity-in-α model to the observed B1 female mortality. The circles show the observed 2-day mortality at each sampled age along with binomial 95% confidence limits. (b) Using the parameter estimates obtained from the best-fit parameters required to fit the line in panel (a), 113,200 (100N) ages at death were randomly generated using the Gompertz equation with gamma-distributed α-values (see the text and the Appendix for details). These ages were used to generate a distribution function for this model and were plotted against the empirical distribution function. If the two distribution functions were identical, they would both fall on the $y = x$ line. Since the empirical curve is above the equality line, especially at higher values of the distribution function, it indicates that B1 females in this cohort are not as long-lived as predicted by the heterogeneity-in-α model that was the best fit to the data of this cohort.

Table 8-1. The fraction of 100 Kolmogorov-Smirnov tests that resulted in significant results at $p < 0.001$

CO			ACO		
Population	Females	Males	Population	Females	Males
CO1	1	1	ACO1	1	1
CO2	1	1	ACO2	1	1
CO3	1	0	ACO3	0	1
CO4	1	0	ACO4	0.09	1
CO5	1	0	ACO5	1	0.99
O			B		
Population	Female	Male	Population	Female	Male
O1	1	1	B1	1	1
O2	1	1	B2	1	1
O3	0.97	1	B3	1	1
O4	1	1	B4	1	1
O5	1	1	B5	1	1

Thirty-five of the forty tests in Tables 8-2 and 8-3 are significant failures at the 5% level. Even if we control for multiple testing using the Bonferroni inequality, there are still 35 significant test results (assessing each individual test using a significance threshold set at $p = 0.00125$). In other words, the probability of flies surviving long enough to reach late life is significantly less than predicted by the heterogeneity-in-α model when it is specifically fit to the data from these cohorts. Thus, for these *Drosophila* data, we can confidently reject the heterogeneity-in-α model as an adequate explanation of mortality in late life.

We have also done a similar analysis for data from the Mediterranean fruit fly, *Ceratitis capitata* (Carey 1993). Like *Drosophila,* medflies show far too few long-lived individuals based on the predictions of the heterogeneity-in-α model, as shown in Figures 8-2 and 8-3. Under the heterogeneity-in-α model, 9.7% of females and 9.3% of males should live to 48 days or longer; in fact, only 1.4% of females and 0.98% of males live this long in the actual cohorts. These discrepancies from the predictions of the heterogeneity-in-α model are statistically significant ($p = 2 \times 10^{-16}$), and therefore, do not lend support to the model.

In summary, the heterogeneity models that best fit the overall mortality patterns of well-studied large cohorts of laboratory organisms did not accurately

Table 8-2. The observed ($\bar{p}$) and expected ($\hat{p}$) probabilities of females living longer than a critical age, with the expectations derived from the heterogeneity-in-α model. The critical ages were: 57.8 (CO), 41.5 (ACO), 39.3 (B) and 80.8 (O)

Population	$\bar{p}$	$\hat{p}$	Prob $(\bar{p} < \hat{p})$	Population	$\bar{p}$	$\hat{p}$	Prob $((\bar{p} < \hat{p}))$
CO1	0.057	0.12	2×10^{-16}	ACO1	0.0089	0.056	2×10^{-16}
CO2	0.027	0.057	1×10^{-13}	ACO2	0.0014	0.10	2×10^{-16}
CO3	0.24	0.32	2×10^{-15}	ACO3	0.071	0.018	1
CO4	0.092	0.13	8×10^{-11}	ACO4	0.01	0.05	2×10^{-16}
CO5	0.021	0.086	2×10^{-16}	ACO5	0.0055	0.11	2×10^{-16}
O1	0.03	0.056	1×10^{-8}	B1	0.026	0.099	2×10^{-16}
O2	0.012	0.026	2×10^{-5}	B2	0.033	0.13	2×10^{-16}
O3	0.017	0.038	5×10^{-8}	B3	0.02	0.076	2×10^{-14}
O4	0.032	0.12	2×10^{-16}	B4	0.055	0.14	2×10^{-16}
O5	0.035	0.057	8×10^{-7}	B5	0.022	0.046	4×10^{-5}

Table 8-3. The observed ($\bar{p}$) and expected ($\hat{p}$ assuming the heterogeneity-in-α model) probabilities of males living greater than a critical age. The critical ages were: 57.8 (CO), 41.5 (ACO), 39.3 (B) and 80.8 (O).

Population	$\bar{p}$	$\hat{p}$	Prob $(\bar{p} < \hat{p})$	Population	$\bar{p}$	$\hat{p}$	Prob $(\bar{p} < \hat{p})$
CO1	0.17	0.21	2×10^{-8}	ACO1	0.68	0.67	0.85
CO2	0.11	0.21	2×10^{-16}	ACO2	0.026	0.10	2×10^{-16}
CO3	0.32	0.085	1	ACO3	0.13	0.26	2×10^{-16}
CO4	0.12	0.05	1	ACO4	0.057	0.17	2×10^{-16}
CO5	0.065	0.0009	1	ACO5	0.024	0.056	2×10^{-16}
O1	0.04	0.13	2×10^{-16}	B1	0.0066	0.11	2×10^{-16}
O2	0.017	0.077	2×10^{-16}	B2	0.012	0.057	9×10^{-14}
O3	0.065	0.16	2×10^{-16}	B3	0.0041	0.036	3×10^{-11}
O4	0.082	0.15	2×10^{-16}	B4	0.02	0.14	2×10^{-16}
O5	0.095	0.19	2×10^{-16}	B5	0.033	0.14	2×10^{-16}

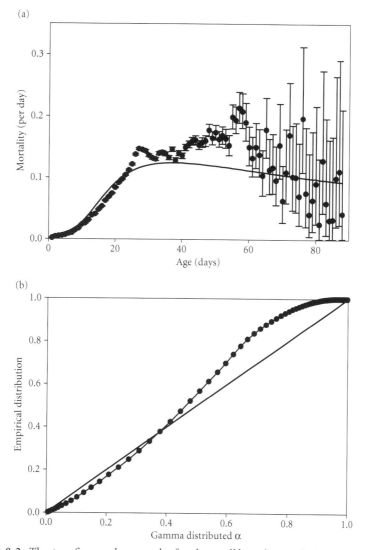

Figure 8-2. The two figures show results for the medlfy male population (Carey 1993, Appendix 2). (a) The line is the weighted least-squares nonlinear fit of the heterogeneity-in-α model to the observed medlfy male mortality. The circles show the observed daily mortality at each sampled age along with binomial 95% confidence limits. (b) Using the parameter estimates obtained from the calculated results shown in (a), 598,118 ages at death were randomly generated using the Gompertz equation with gamma-distributed α-values (see text for details). These simulated deaths were then used to generate a distribution function for this model, which was plotted against the empirical distribution function. If the two distribution functions were identical, they would fall on the $y = x$ line. Since the resulting curve is above the equality line, especially at higher values of the distribution function, it indicates that medfly males are not as long-lived as predicted by the heterogeneity-in-α model.

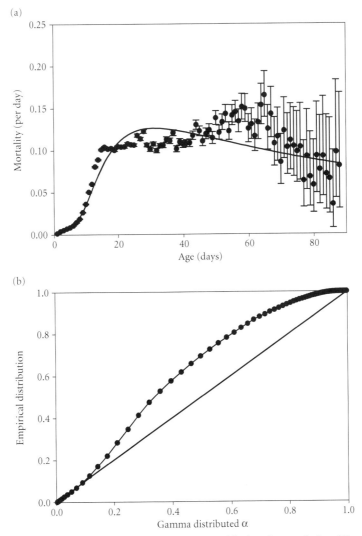

Figure 8-3. The two figures show results for the medfly female population (Carey 1993, Appendix 3). (a) The line is the weighted least-squares nonlinear fit of the heterogeneity-in-α model to the observed medlfy female mortality. The circles show the observed daily mortality at each sampled age along with binomial 95% confidence limits. (b) Using the parameter estimates obtained in (a), 605,528 ages at death were randomly generated using the Gompertz equation with gamma-distributed α-values (see text for details). These were used to generate a distribution function for this model and plotted against the empirical distribution function. If the two distribution functions were identical, they would fall on the $y = x$ line. Since the curve is above the equality line, especially at higher values of the distribution function, it indicates that medfly females are not as long-lived as predicted by the heterogeneity-in-α model.

predict the age at death of the last fly to die in the cohort. Furthermore, these heterogeneity models predicted that more flies would be alive at late ages than were actually observed. Mueller et al. (2003) also showed that the variance in mortality rates changed little with age in laboratory *Drosophila*, leaving aside very early and late ages, which is contrary to predictions that have been made based on the lifelong heterogeneity model (cf. Service 2004). Collectively, these experimental results do not support a key role of demographic heterogeneity in late-life mortality. But they do not necessarily exclude a contribution of lifelong heterogeneity to some of the slowing in mortality rates at late ages. However, it is unlikely that this theory can explain late-life patterns entirely on its own.

TESTING LIFELONG HETEROGENEITY THEORIES BY MANIPULATING ROBUSTNESS

When individual robustness is radically improved by selection for increased robustness, and it is hypothesized that lifelong heterogeneity is the cause of late-life mortality-rate plateaus, then late-life mortality-rate plateaus should change with respect to their timing. The more robust population will be affected by environmental variation in a radically different fashion than the less robust population, and therefore the late-life characteristics of the two populations ought to be very different. In reasoning like this, we are implicitly accepting the presupposition that robustness at one adult age is strongly correlated with robustness at all adult ages. In doing so, we are conforming to dictates of lifelong heterogeneity theory as a bare supposition, not arguing that this is in fact the case.

Drapeau et al. (2000) tested this lifelong heterogeneity prediction using populations of *Drosophila* selected for starvation resistance (SO) and comparing them to their controls (CO and RSO); they found no differences in late life (Figure 8-4). By contrast, a post hoc reanalysis of these data by Steinsaltz (2005) led him to different conclusions. It is important to note, however, that in a major methodological departure, Steinsaltz chose to remove from his analysis the observed mortality data in early life. With such a selective omission, it is hardly surprising that the results might differ. The process of removing data is always fraught with danger, because it is by and large a subjective procedure often guided by a priori expectations that are in fact part of the hypotheses being tested. Therefore, we believe that we can reasonably conclude that selectively produced differentiation in robustness does not consistently affect the presence of late-life mortality-rate plateaus, contrary to the line of reasoning outlined above that was based on the lifelong heterogeneity hypothesis.

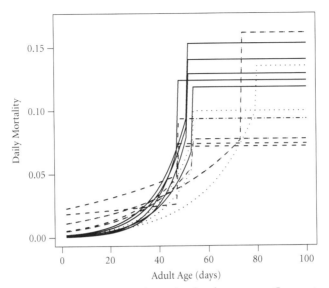

Figure 8-4. The daily mortality rates from the fitted two-stage Gompertz model for females from the SO (solid lines), CO (dashed lines), and RSO (dotted lines) populations. An analysis of the breakday and plateau height showed no significant differences (Drapeau et al., 2000).

Carey has argued (Carey et al. 1995; Carey 2003) that if mortality is increased by increasing the population density, then the age at which a mortality plateau occurs should decline. This is a robustness-reducing environmental manipulation, the converse of the experimental strategy of Drapeau et al. (2000). This follows because at high density the less robust groups are eliminated faster, and thus the age at which only the most robust groups are left (or the breakday, in our terminology) should come sooner. However, in experiments with Mediterranean fruit flies, changing adult density had no detectable effect on the age at which mortality rates leveled off (Carey et al. 1995; Carey 2003). From these results, Carey (2003, p. 97) concluded that the "leveling off of mortality is not an artifact of changes in cohort composition."

TEST OF LIFELONG HETEROGENEITY THEORY USING CORRELATIONS OF AGE-SPECIFIC FECUNDITY

As we discussed in Chapters 2 and 4, previous studies have found that fecundity, like mortality rates, plateaus at late ages in several independent cohorts of *D. melanogaster* (Rauser et al. 2003, 2005b, 2006b). Although evolutionary theory based on the age-specific decline in the force of natural selection can explain the decline and plateau in fecundity at late ages (Hamilton 1966; Rauser et al. 2006a), Rauser et al. (2005a) thought it worthwhile to consider the possibility that lifelong heterogeneity in individual female fecundity could cause a spurious plateau in average late-life fecundity.

Although Vaupelian theory has not been extended to include fecundity by its original proponents, several post hoc explanations that are based on demographic heterogeneity might be able to explain the existence of late-life plateaus in fecundity, as we outlined in Chapter 6. Fecundity models analogous to the Vaupelian model for mortality could be based on lifelong differences in individual female fecundity. One possible heterogeneity-based explanation analogous to the Vaupelian model for mortality for observing late-life fecundity plateaus in cohorts of *D. melanogaster* is that females that lay a large number of eggs die prematurely, leaving only the females that always laid a small number of eggs preponderant among late age groups. Another possibility is that some females both live longer and sustain fecundity better. In either case, if fecundity plateaus are a consequence of lifelong differences in egg laying, then measuring individual fecundity patterns for females comprising a large cohort, and comparing the fecundity of individuals that live to lay eggs in late life with those that do not, would test *either* possibility, and we did just that (Rauser et al. 2005a).

The first *lifelong heterogeneity in fecundity* hypothesis described above is implicitly based on a trade-off between egg laying and lifelong robustness, while the second is a generalization of the Vaupelian lifelong-robustness theory from mortality to all age-specific life-history characters (cf. Vaupel et al. 1979). Furthermore, many other variations on these themes are conceivable. However, regardless of the numerous conceivable lifelong heterogeneity in fecundity hypotheses, all of them have in common the ability to infer late-life fecundity patterns from attributes of young individuals in a cohort, just as demographic theories of late-life mortality hypothesize that mortality rates plateau because of individual heterogeneity effects that are present throughout life (vid. Vaupel et al. 1998). To be specific, lifelong heterogeneity theories of mortality assume that individuals are imbued with lifelong consistent levels of robustness that define their mortality rates. As a result, individuals within a cohort that are less robust *throughout life* die at earlier ages, leaving individuals with lifelong superiority in robustness predominant in the cohort at later ages, causing a slowing of mortality rates (Vaupel et al. 1979; Vaupel 1988, 1990; Pletcher and Curtsinger 2000).

A major problem in testing lifelong heterogeneity theories with regard to mortality is that an individual's rate of aging with respect to mortality cannot be measured readily while it remains alive, so lifelong heterogeneity for robustness has only been studied indirectly where mortality is concerned. However, with fecundity this is not the case, as individual age-specific fecundity over a lifetime can easily be measured within a cohort. Thus, fecundity can be used to test the *general* concept of lifelong demographic heterogeneity (Rauser et al. 2005a), because average population fecundity shows the same plateauing pattern at late ages as mortality rates.

Other studies of individual fecundity trajectories helped to motivate this experimental strategy. For example, Müller et al. (2001) looked at fecundity and death patterns in medflies and found no apparent trade-off between reproductive output and lifespan. This is preliminary evidence against one version of the lifelong heterogeneity in fecundity theory, specifically the hypothesis that females that lay a large number of eggs should die at earlier ages. In another study by Novoseltsev et al. (2004), flies with short lifespans did not have higher mean fecundity during their midlife "plateau"—note that their usage of this term does not correspond to ours—compared to flies that lived a medium number of days. This result is also inconsistent with the predictions of the first type of heterogeneity theory for fecundity adduced above. However, they did show that the longest-lived flies had a lower mean fecundity than the medium- and short-lived flies, though this difference was not always statistically significant. Overall, at the time we decided to test the lifelong heterogeneity in fecundity hypothesis using lifetime fecundity trajectories for individual females within a cohort, it was not clear from the published literature whether any form of this theory was most likely to be correct.

We tested whether *observable* lifelong heterogeneity in fecundity can be used to predict the properties of the late life of individual flies, including the survival of individual flies to the late-life period. This was done by measuring for individual females both daily fecundity over the entire lifetime and the age of death, and then testing whether the age-specific fecundity of females that lived to lay eggs at late ages differed significantly throughout life from the age-specific fecundity of females that died before the onset of the cohort's plateau in fecundity (Rauser et al. 2005a). Over the course of this experiment, we counted 3,169,101 eggs laid over the lifetime of 2,828 females.

Our study used the outbred laboratory-selected CO_1 population of *D. melanogaster* selected for midlife reproduction, as described in Chapters 4 and 5 (see Figure 4-2). These populations are cultured using females 28 days of age (Rose et al. 1992), and at the time of the study that we describe here, they had been maintained at population sizes of at least 1,000 individuals for more then 170 generations. Both late-life mortality-rate plateaus and late-life fecundity plateaus have been observed in the CO populations (Rose et al. 2002; Rauser et al. 2006b). Three separate assays were performed using large cohorts from the CO_1 population (see Rauser et al. 2005a for experimental details).

The basic flavor of our results can be illustrated by examining the third experiment (Figure 8-5). We have plotted the lifetime fecundity of each female with a line of different shades of white and black (Figure 8-5). A line in this figure ends when the fly dies. It is clear from this figure that, just prior to death, fecundity has declined dramatically relative to that of the other females that are still alive (this is the death spiral described in more detail in Chapter 9). More importantly, if we look at female fecundity early in life (ages 12–25 days), we see very little difference between the flies that die first and those that live much longer. These visual impressions are confirmed by more formal statistical tests described in Rauser et al. (2005a).

Our test of lifelong heterogeneity theories for fecundity based on these data depends on whether late-life fecundity or survival is predictable from differences in egg laying between individual females at earlier phases of adult life, including early adulthood. For example, a cohort that shows lifelong heterogeneity in egg laying with strong trade-offs between reproduction and survival should have females that consistently lay more eggs quickly and then die at earlier ages, leaving only females who have always laid eggs at a low rate preponderant at late ages. Alternatively, a cohort with some members that show lifelong superiority with respect to all adult life-history characters, including all age-specific survival probabilities and all age-specific fecundities, should allow us to predict survival to late

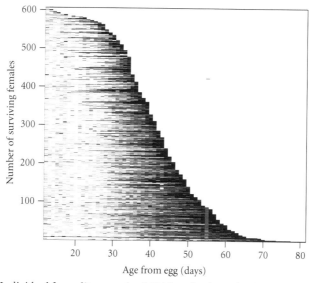

Figure 8-5. Individual fecundity records of 606 females from the third assay. Females were rank-ordered by age of death within this cohort on the *y*-axes, and the individual age-specific fecundity patterns of each female were plotted horizontally on each graph along the *x*-axes. Female fecundity was divided into five categories and color-coded accordingly: 0 eggs, 1–9 eggs, 10–19 eggs, 20–49 eggs, 50–194 eggs. The zero-fecundity category is black, and the shades get progressively lighter as the number of eggs increases.

life from early fecundity data. However, our analysis showed that neither of these hypotheses is likely to be correct, because early-life fecundity did not predict late-life characteristics. The data suggest that there is a significant amount of age-specific variation in fecundity, but that it has no predictive value until 12–15 days after the start of reproduction (Rauser et al. 2005a).

Similarly, our simple model for heterogeneity in fecundity required a 15-fold difference in fecundity between the high and low egg layers in order to simulate accurately our observed cohort fecundity values. It would be interesting to see if average cohort fecundity plateaus at late ages in genetically homogeneous cohorts. A plateau in fecundity under these circumstances would indicate whether age-specific, though not lifelong, genetic heterogeneity plays a role in late-life fecundity patterns, because it would not eliminate the contribution of age-specific environmental heterogeneity. However, it is unlikely that exogenous environmental heterogeneity has much of an effect on the existence of fecundity plateaus, as we have observed fecundity plateaus under both constant and varying environmental conditions (Rauser et al. 2005a, 2005b, 2006b).

Other studies of the fecundity trajectories of individual flies generally support our experimental findings and do not support the predictions of lifelong heterogeneity models for fecundity. As we explained above, Müller et al. (2001) found no apparent trade-off between reproductive output and lifespan in medflies, which is additional evidence against the type of model that we simulated here. A lifelong heterogeneity model for fecundity with strong trade-offs between reproduction and survival predicts just the opposite: females that lay a large number of eggs should die at earlier ages, which is equivalent to a trade-off between reproduction and lifespan. Analysis of the phenotypic relationship between lifetime reproduction and lifespan in our flies indicates that long life is also coupled with increased lifetime reproduction (Rauser et al. 2005a). Furthermore, as already mentioned, Novoseltsev et al. (2004) showed that flies with short lifespans do not have higher mean fecundity during what they call their midlife "plateau" compared with flies that live a medium number of days. This too is not consistent with the predictions of the lifelong trade-off heterogeneity theory for fecundity. (Note that their "plateau" is a midlife plateau for individual females, while our "plateau" usage refers to average population fecundity at very late ages.) However, they did show that the longest-lived flies had lower mean fecundity than the medium- and short-lived flies, but not always significantly lower. An analysis of the relationship between the mean number of eggs each female laid per day and lifespan in our flies suggests a similar relationship. That is, longer-lived flies had a slightly lower mean number of eggs laid per day (Rauser et al. 2005a), but not low enough compared to shorter-lived flies to significantly improve our ability to predict which females would be long-lived plateau females, or not, at earlier ages.

Our studies of fecundity indicate that there is significant, predictive, age-specific heterogeneity in fecundity within large cohorts, which is to be expected in a genetically heterogeneous outbred population. This heterogeneity is not lifelong, nor is it sufficient to cause late-life plateaus in average population fecundity. The most significant type of age-specific heterogeneity was between flies about to die versus those that were not about to die. Because lifelong heterogeneity in fecundity hypotheses are based on the same types of underlying assumptions as lifelong heterogeneity theories proposed to explain late-life mortality-rate plateaus, our test of such lifelong models for fecundity is relevant to the mortality models as well. If lifelong heterogeneity effects are generally related to late life, then they should have passed this test. Our results refute at least one general class of heterogeneity theories, those based on fixed *lifelong* differences in fecundity.

CONCLUSION: EVIDENCE AGAINST LIFELONG HETEROGENEITY THEORIES OF LATE LIFE

A great deal of experimental effort has been devoted to tests of lifelong heterogeneity as an explanation of late-life phenomena involving both mortality and fecundity. It should be pointed out that, while we have not been exponents of lifelong heterogeneity theory for some time, others who have conducted tests of it apparently were, at least prior to collecting their data (e.g., Khazaeli et al. 1998). But regardless of the views of the experimenters, no one has found support for this hypothesis in critical experimental tests, as opposed to post hoc fits of demographic data using the lifelong heterogeneity hypothesis, whether merely conjectural (e.g., Greenwood and Irwin 1939; Beard 1964; Vaupel et al. 1979) or quantitative and specific (e.g., Kowald and Kirkwood 1993). Instead, there is a proliferation of evidence against lifelong heterogeneity theories of late life.

In fairness, however, it should be pointed out that there is an abundance of alternative mortality models in demography, and some of the tests described here, such as those of Mueller et al. (2003), are model-dependent. Choosing other demographic models might lead to different results, and we have not repeated our analyses over the full range of published, much less conceivable, demographic models. Naturally, this would be an endless enterprise.

On the other hand, some of the tests of lifelong heterogeneity that we have discussed in this chapter are not particularly model-dependent. Examples of relatively model-independent experimental tests are those of Fukui et al. (1993), Khazaeli et al. (1998), Drapeau et al. (2000), and Rauser et al. (2005a). Lifelong heterogeneity theory has only received falsifications in these tests, never corroborations, when the experimental results have been clear. However, we have already encountered some maneuvering with respect to these ostensible falsifications (see Mueller and Rose 2004) as we have discussed in this chapter with reference to the reanalysis of our data by Steinsaltz (2005). We can expect more such challenges to experimental refutations of lifelong heterogeneity, since the theory is mathematically elegant as well as intuitively attractive for those who cannot accept the conclusion that aging could possibly cease at the level of individuals.

Thus, like those who espouse special creation or intelligent design against the evidence accumulated against these hypotheses, we can expect devotees of life-long heterogeneity theories for late life to be with us for some time. But even if they are wrong, they serve the useful role of devil's advocates against the claims of Hamiltonian aging research, keeping its proponents on their toes.

Death Spirals

Unlike the case of mortality, there are no general-purpose models to describe age-specific fecundity patterns. Here we show that combining the effects of natural selection on fecundity with a newly discovered physiological phenomenon we call the death spiral results in highly accurate descriptions of female fecundity in Drosophila.

MODELS OF FECUNDITY

Unlike the case of the demography of survival, there is no universal simple model that describes age-specific fecundity. Most models of the patterns of age-specific fecundity that are used in biological research are simple phenomenological models. These models have as their primary goal the accurate prediction and statistical analysis of fecundity (e.g., Geyer et al. 2007; Shaw et al. 2008). While such descriptive models serve an important role in population biological research, another important goal of modeling is to understand the forces shaping age-specific fecundity. For us, an important force is the impact of natural selection on patterns of age-specific fecundity.

There have been previous attempts to develop models of age-specific fecundity based on physiological and evolutionary forces. For instance, Novoseltsev et al. (2003, 2004) proposed a model that assumes a decline in fecundity late in life due to age-related oxidative vulnerabilities. While this is not an explicitly evolutionary model, one could argue that these vulnerabilities show a decline with age due to the declining force of natural selection. Cichon (2001) and Shanley and Kirkwood (2000) developed evolutionary models for life history, including fecundity, following the optimal life-history paradigm pioneered by Gadgil and Bossert (1970). These theories can incorporate a number of complex effects on fecundity, although they often ultimately rely on the questionable assumption that natural selection maximizes the lifetime number of offspring.

In Chapter 4 we developed Hamiltonian evolutionary theory that suggests some very general patterns for age-specific fecundity. We elaborate on that general model here, with the addition of a newly discovered physiological phenomenon called the *death spiral* that profoundly affects fecundity just prior to death. The resulting model is relatively simple, biologically motivated, and provides accurate descriptions of age-specific fecundity in *Drosophila*. This makes it a potential candidate model for the demography of fecundity in other species as well.

THE DEATH SPIRAL PHENOMENON

There are three important stages of life from the perspective of evolutionary biology (Rose et al. 2006; Shahrestani et al. 2009). The first is the developmental period prior to reproduction. During this stage, natural selection works with maximum efficiency to weed out genetic variants that reduce survival before the onset of reproduction. Any individual that fails to survive this period will have zero fitness in the absence of altruistic interactions with related individuals. The powerful action of natural selection during this stage doesn't guarantee survival, even under optimal conditions, because of recurring mutations, segregational genetic load, and developmental accidents. But it does mean that this stage of life is the primary beneficiary of natural selection for enhanced survival. The aging phase is the second stage of life; it is also the period following the onset of reproduction, which brings fecundity into consideration as a component of fitness. The theory of selection in age-structured populations predicts eventual decreases in age-specific fitness components, even under ideal conditions, including age-specific fecundity, during this second stage (Charlesworth 1994). The third stage of life has been called *late life* (e.g., Rose et al. 2002; Rauser et al. 2006a). Because at advanced ages age-specific selection is effectively absent, our expectation is that age-specific fecundity under protected conditions will plateau.

We have reviewed data suggesting this in Chapter 3. A more formal explanation of fecundity plateaus was then presented in Chapter 4.

But it turns out that fecundity trajectories are even more complicated. In the large-scale study of age-specific patterns of female fecundity in *Drosophila* that we described in Chapter 8, we discovered a fourth life-cycle phenomenon that we have called the *death spiral* (Rauser et al. 2005a; Mueller et al. 2007). For a period of 6–15 days prior to death, the fecundity of females that are about to die drops at a much faster rate than the fecundity of similarly aged females that are not about to die. This result was discovered by comparing the slopes of the line describing fecundity versus age as a function of the prospect of death for individual females. In Chapter 8, we saw that the ability to distinguish between plateau and nonplateau females improves as more nonplateau females are about to die, or enter the death spiral, in our terminology (see Figure 8-5). The death spiral is detectable across a wide range of adult ages (Figure 9-1); it may signal a very general decline in physiological health prior to death. The death spiral has also been independently documented in *D. melanogaster* by other laboratories (e.g., Rogina et al. 2007).

Phenomena similar to the death spiral have been observed in other organisms. Christensen et al. (2008) monitored the physical and cognitive abilities of 2,262 Danish individuals born in 1905. Over the course of the study, the individuals were between 92 and 100 years of age. They found that the physical and cognitive scores of a group of individuals who died within two

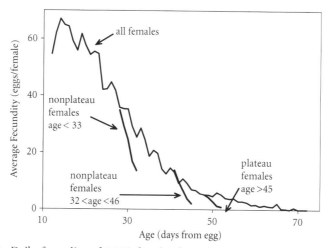

Figure 9-1. Daily fecundity of 1,100 females from the CO1 population (Rauser et al. 2005a; see also Chapter 8). The thin solid line shows the average fecundity for all females. The thick lines show the average fecundity for females 5 days prior to death that have entered the death spiral at different age ranges. It appears that death is accompanied by a dramatic decline in fecundity independent of the age at death.

years of the initial measurements were significantly lower than the scores of similarly aged individuals who did not die. Similarly, male medflies will often be found on their backs prior to death, although if the fly can right itself, it continues to display more or less normal behavior (Papadopoulos et al. 2002). This supine behavior also appears to be a reliable signal of impending death.

There are a host of interesting questions about the process of dying that could be addressed if it were possible to identify reliably the females that have entered the death spiral prior to their actual death. It is reasonable to suppose that, if fecundity is undergoing a dramatic decline prior to death, then other aspects of physiology may also be changing dramatically. Since many physiological assays in *Drosophila* and other species are destructive, it will not always be possible to collect these physiological data immediately prior to the death of a test female. This inability will limit the study of the process of dying.

PREDICTING DEATH IN FEMALE *DROSOPHILA*

We have recently developed statistical methods for predicting whether an individual female is in the death spiral or not (Mueller et al. 2009). These methods were validated using three different data sets that we now describe in more detail.

Our study used the data collected from the lifelong heterogeneity in fecundity experiment described in Chapter 8 that followed the daily fecundity and time of death of 2,828 individual females from the outbred CO_1 laboratory population of *D. melanogaster* (see Rauser et al. 2005a). This population is one of the five replicate CO populations derived in 1989 from five corresponding O populations (Rose 1984b) and is selected for midlife reproduction (age 28 days; Rose et al. 1992), as further described in Chapters 4 and 5 (see Figure 4-2). The data that we analyze here were derived from three large cohorts of flies from the CO_1 replicate population and were collected to test the lifelong heterogeneity in fecundity hypotheses, as described in Chapter 8 (see Rauser et al. 2005a and Mueller et al. 2009 for experimental details).

Suppose we have a cohort of flies aged t days, an age that we will call the *target age*. At the target age, we would like to separate females into two groups: those that are in the death spiral and those that are not. To be more specific, we consider a female to be in the death spiral if she is expected to die on day $t+1$, $t+2, \ldots, t+v$, where the age increment v is the maximum length of the death process. Based on our previous estimates of the duration of the death spiral in *Drosophila*, v could range from 5 to 14 days for a female who enters the death spiral at day t.

Since it is more likely that female flies well into the death spiral will exhibit altered physiology compared to females that have just begun the death spiral, we have set $v = 5$ for the data analysis that follows. We regard this as a conservative assumption that also allows the female to be in the death spiral for several days prior to the target day, and her fecundity should reflect this. This definition of the duration of the death spiral means that it is possible that some females that have started their death spiral might be mistakenly classified as non-death-spiral females because they die at an age $> t+5$. However, it is much less likely that a female who would die within 5 days of the target age would not be classified in the death spiral.

In the absence of any information about female fecundity, we could still use the survival of flies prior to the target day to estimate the chance of a fly dying over the next 5 days. We expect that experiments designed to measure the physiology of death spiral females would collect flies at ages well before a mortality plateau, in which case survival might be accurately predicted by the Gompertz equation (Gompertz 1825; Mueller et al. 1995). Under this model, the chance of dying in the 5 days following the target age (P) would be

$$P = 1 - \frac{p_{t+5}}{p_t}, \tag{9-1}$$

where p_t is the chance of surviving to age t. The probability of surviving to age t is given by the Gompertz equations as

$$p_t = \exp\left\{\frac{A\left[1 - \exp\left(\alpha t\right)\right]}{\alpha}\right\}, \tag{9-2}$$

where A is the age-independent Gompertz parameter and α is the age-dependent parameter.

Information from a cohort's survival records allows us to predict with some accuracy how many females should be in the death spiral. However, with this information alone, the only way to use the information would be to randomly choose the appropriate number of females for each group, that is, those in the death spiral and those not in the death spiral. But data on age-specific female fecundity might, in principle, give better information for making more precise predictions about which females to put in each category. As shown in Mueller et al. (2009), the total number of eggs laid by females 3 days prior to an assay can be used to reliably classify females as either in the death spiral or not. Thus, a practical protocol for creating the two groups of females can make use of demographic predictions from both the observed number of deaths prior to the assay and age-specific female fecundity (Figure 9-2).

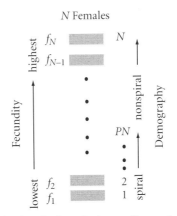

Figure 9-2. A collection of N females is ranked according to their 3-day fecundity, with f_1 being the lowest value of fecundity and f_N the highest. From the survival data, the parameters of the Gompertz equation can be estimated and then used in Equation 9-1 to predict P, the fraction of the population that should be in the death spiral. Thus, the PN females with the lowest fecundity values form the death-spiral group and the remaining females form the non-death-spiral group. Techniques for improving this classification scheme are discussed in Mueller et al. (2009).

AN EVOLUTIONARY HETEROGENEITY MODEL OF FECUNDITY

We will now present a statistical model of late-life fecundity by distinguishing between the egg laying of females before and during their death spiral. Since this model uses predictions from evolutionary theory and the fact that females are heterogeneous (e.g., *spiral vs. nonspiral*), we call this model the *Evolutionary Heterogeneity Fecundity model*, or the *EHF model* for short. The basic pattern of female age-specific fecundity before the death spiral that is assumed in our model is that in middle to late life, fecundity shows a roughly linear decline until the fecundity breakday (*fbd*), after which fecundity remains constant (Figure 9-3). These assumptions lead to the following relationship between age (t) and fecundity ($f(t)$):

$$f(t) = \begin{cases} c_1 + c_2 t, & \text{if } t \le fbdf \\ c_1 + c_2 fbd, & \text{if } t > fbd \end{cases} \tag{9-3}$$

Just before death, during the death spiral, it is assumed that fecundity declines at a more rapid rate (Figure 9-3). If the duration of the death spiral is w days and a particular female dies at age d, then her fecundity for w days prior to death, $\tilde{f}(t)$, is given by

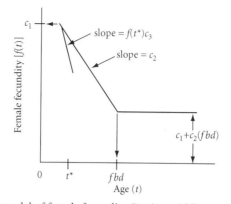

Figure 9-3. An EHF model of female fecundity. During middle ages, the decline in female fecundity at age t is described by the line $f(t) = c_1 + c_2t$. At age *fbd*, called the *fecundity breakday*, female fecundity reaches a plateau of $c_1 + c_2$*fbd* eggs per day. Females about to die are assumed to enter a death spiral that involves a steep decline in fecundity. If a female begins this death spiral at age t^*, then it is assumed that fecundity declines linearly from that age until death with a slope of $c_3f(t^*)$. This slope may be the same for all flies or may vary for pre- and postplateau females. The duration of the death spiral is assumed to be of fixed length. It may be estimated independently from data or via regression from the population fecundity data.

$$\tilde{f}(t) = f(d-w) + f(d-w)c_3(w+t-d). \qquad (9\text{-}4)$$

This formulation of the death spiral assumes that the slope of the decline is proportional to the average fecundity of females at the age at which the death spiral begins. Both $f(t)$ and $\tilde{f}(t)$ are constrained to have nonnegative values. Accordingly, the complete four-parameter model for age-specific fecundity with parameters $\theta = (c_1, c_2, c_3, fbd)$ is

$$F(t,d,\theta) = \begin{cases} f(t) \text{ if } t < d-w \\ \tilde{f}(t) \text{ otherwise} \end{cases}. \qquad (9\text{-}5)$$

An important parameter of this stochastic EHF model is the duration of the death spiral, w. While the value of this parameter could be estimated from a regression analysis, we first examined individual fecundity patterns to see what an empirically estimated value of w might be. To accomplish this, we analyzed the individual fecundity data collected for the $CO_{1\text{-}1}$ population described in Chapter 8 and in Rauser et al. (2005a) as follows.

We separated all females within the cohort into two groups: those dying before the breakday (*fbd*) and those dying after *fbd*. The age of these flies was then rescaled to the number of days before their death rather than to their

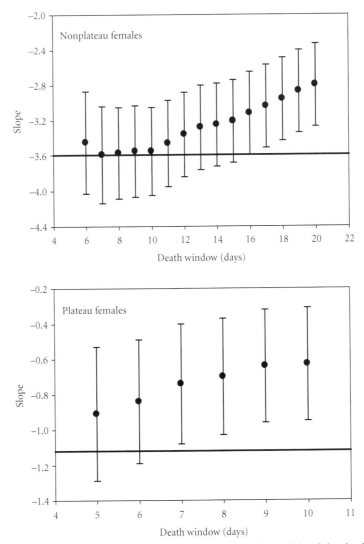

Figure 9-4. The slope of the death spiral as a function of the width of the death spiral window, in days, for nonplateau and plateau females. The horizontal line shows the base slope of the death spiral using only 5 (nonplateau females) or 4 (plateau females) days of fecundity observations before death. Each point represents the slope with additional observations added. The error bars can be used to determine when these slopes are significantly different from the base slope. The error bars are twice the square root of the variance of the sum of the two estimated slopes (the base slope and the current slope).

absolute age. From these data, we then estimated the slope of female fecundity over the days before death using different numbers of observations, varying the duration of the death spiral. Our expectation was that, as we added observations farther back in time from the day of death, the fecundity value should return to the average cohort fecundity value, causing the magnitude of the slope of the fecundity decline to fall relative to its value when only the first few days before death are used to estimate this slope. This analysis showed that the slope remains unchanged for nonplateau females until 16 days before death, suggesting a death-spiral duration of 15 days (Figure 9-4). In plateau females, the slope change occurs at 7 days before death, suggesting a death window of 6 days. Based on these results, we used a death-spiral duration of 10 days in models that treat w as a fixed constant for the sake of simplicity.

Our basic model of female fecundity (Equation 9-5) has four parameters, $\theta = (c_1, c_2, fbd, c_3)$, but we also examined three other variants of this model. We studied a five-parameter model, which assumes that the slope of the death spiral, c_3, may be different for preplateau and postplateau females. We also generalized this five-parameter model to a six-parameter model by making the *duration* of the death spiral a model parameter. The fourth model that we considered was the most complicated; it was the same as the six-parameter model, except that the duration of the death spiral was allowed to differ for flies dying before the plateau and after the plateau.

Estimating the parameters of these EHF models requires information on both age-specific fecundity and mortality. Without the mortality data, we cannot directly infer the timing of female death spirals. Therefore, we identify three classes of experimental data that we have been able to analyze. (i) First, there are experiments that have measured fecundity on individual females and have also recorded the age at death of these females (specifically, data from Rauser et al. 2005a). These are the best data and permit direct estimates of model parameters. (ii) Second, there are experiments where the number of deaths of a cohort of females is recorded at regular time intervals, but fecundity is observed on groups of females, not individuals. (iii) Finally, there are experiments where fecundity was observed on groups of females, but no survival data were recorded. We discuss the results and methods of analysis for each of these three classes of experimental data in turn.

Individual Fecundity and Survival Records

Let the observed number of eggs laid by a female i at age x be, f_{ix}, $i = 1, \ldots, N$ and $x = t_b \ldots t_d$. Thus, t_b is the age from egg at which female reproduction begins and $t_d + 1$ is the greatest age at death of the N females. For each of the N females, let the observed age at death be d_i. With these observations, we can compute the average fecundity at each age from

$$f_x = \frac{1}{n_x} \sum_{\substack{i \text{ such that} \\ d_i > x}} f_{ix} \qquad (9\text{-}6)$$

based on records of n_x females still alive at age x. The predicted average fecundity $(F_x(\hat{\theta}))$ at age x for parameter values $\hat{\theta}$ is calculated as

$$\frac{1}{n_x} \sum_{\substack{i \text{ such that} \\ d_i > x}} F(x, d_i, \hat{\theta}), \qquad (9\text{-}7)$$

where $F(x, d_i, \hat{\theta},)$ is one of the fecundity models such as Equation 9-5. The model parameters, $\hat{\theta}$, are then chosen to minimize

$$\frac{1}{(t_d - t_b + 1)} \sum_{x=t_b}^{t_d} \left[f_x - F_x(\hat{\theta}) \right]^2. \qquad (9\text{-}8)$$

Since there are so many more females at the young ages, we have chosen a least squares statistic that treats each age as an equivalent sampling unit. However, since there are fewer observations at the older ages, we expect these predictions to be less reliable. This uncertainty will be reflected in the size of the confidence intervals we compute with these regression predictions.

To evaluate the uncertainty in the predicted values of female fecundity, we used bootstrap resampling of our data. A bootstrap sample, $\tilde{f}_{ix}$, was generated by taking a sample of N females with replacement from the original set of N females. This sampling also produced N bootstrap ages at death, $\tilde{d}_i$. With this bootstrap sample, we utilized the methods summarized in Equations 9-6 to 9-8 to obtain a least squares estimate, $\tilde{\theta}$. The parameter $\tilde{\theta}$ was then used to predict the mean fecundity at each age

$$\tilde{F}_x(\tilde{\theta}) = \frac{1}{\tilde{n}_x} \sum_{\substack{i \text{ such that} \\ \tilde{d}_i > x}} F(x, \tilde{d}_i, \tilde{\theta}),$$

where $\tilde{n}_x$ is the number of females alive at age x in the bootstrap sample. One hundred bootstrap samples were generated, and 96% confidence bands on the average value of the 100 $\tilde{F}_x(\tilde{\theta})$ were derived from the second smallest and 99th largest values of $\tilde{F}_x(\tilde{\theta})$.

We performed an analysis of the EHF models using the individual fecundity and mortality data collected by Rauser et al. (2005a) and found that the four-parameter model (Equation 9-5) was most often the best model over all three indices used for assessing model fit (see the Appendix section for this chapter for statistical details of this analysis, including Table A9-1). This result, combined with the normal scientific preference for the most simple

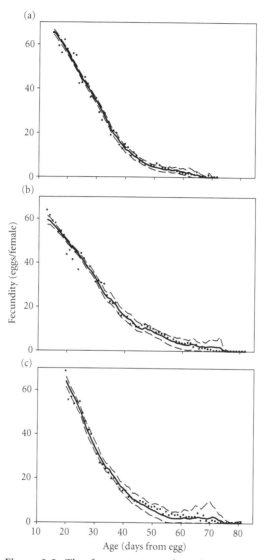

Figure 9-5. The four-parameter fecundity model (dark solid line) and 96% confidence interval (light black lines) for cohorts (a) CO_{1-1}, (b) CO_{1-2}, and (c) CO_{1-3}. The circles are the mean observed fecundity at each age.

model, suggests that Equation 9-5 is perhaps the best and simplest description of age-specific female fecundity. We have used this four-parameter model to compare the average predicted fecundity from the model with the observed average fecundity in the three CO_{1-1}, CO_{1-2}, and CO_{1-3} cohorts (Figure 9-5).

Although the fit of the four-parameter model is very good, we do not consider goodness of fit alone to be the sole important criterion for assessing the utility of this type of model. This model is our hypothesis about the evolutionary forces molding age-specific fecundity as well as a reflection of individual physiological decline prior to death. Thus our belief in this model, or any other model, cannot be evaluated solely by its goodness of fit. The strongest virtues of this model are the soundness of its theoretical underpinnings and its ability to predict new experimental observations, which are in part addressed by the cross-validation statistic.

Individual Survival Records and Group Fecundity Records

To estimate the basic four-parameter model's (Equation 9-5) coefficients and to provide confidence intervals about the estimated parameter values, we compared observed fecundity values with those derived from simulations. The simulations generated ages at death from the two-stage Gompertz mortality model. The parameter estimates for the two-stage Gompertz model were obtained in independent mortality experiments.

Our experimental data (described in Rauser et al. 2006b) consisted of an initial cohort of 3,200 females. These females were maintained in vials with four females per vial. At each age, if there were more than 400 surviving females, a sample of 100 vials was chosen to estimate fecundity. Once the number of survivors dipped below 400, all vials were used to estimate fecundity. Thus, the per-capita fecundity of females in vial i at age x is given by $f_i(x)$, $i = 1, 2,\ldots, n_x$, where n_x is the number of vials used to estimate fecundity at age x. Age-specific fecundity estimates started at age t_b, which was 30 days from egg for all five populations, and ended at day t_d, the last day there were four live females, which varied among populations.

In our numerical analysis, the bootstrap fecundity sample at age x was generated by taking n_x samples with replacement from $f_i(x)$. This bootstrap sample is represented as $\tilde{f}_i(x)$, $i = 1, 2,\ldots, n_x$.

The independent mortality data were used to estimate the parameters of the two-stage Gompertz that were used in simulations of mortality. The distribution function of the two-stage Gompertz, $G(x)$ is

$$\exp\left\{\frac{A\left[1-\exp(\alpha x)\right]}{\alpha}\right\} \textit{if } x \leq mbd$$

$$\exp\left\{\frac{A\left[1-\exp(\alpha x)\right]}{\alpha}\right\}\exp\left[\tilde{A}(mbd-x)\right]\textit{if } x > mbd$$

The age at death, d, for 3,200 females in the bootstrap sample was simulated by the inverse transform algorithm as $d = G^{-1}(U)$, where U is a uniform random number on the interval $(0,1)$ (Fishman 1996). At each age we took a sample of 400 females or, if there were fewer than 400 survivors, all females were used. Let the number of females used at each age be $\tilde{n}_x$. With the simulated age at death for these females and an estimate of the model parameter θ_0, we estimated the predicted fecundity of each female as $F(x, d_i, \theta_0)$ (Equation 9-5) for $i = 1, \ldots, 3,200$. The bootstrap estimate of the average fecundity at age x for parameter θ_0 was then estimated from the average, $\tilde{F}_x(\theta_0) = \dfrac{1}{\tilde{n}_t} \displaystyle\sum_{i=1}^{i=\tilde{n}_t} F(x, d_i, \theta_0)$. The least-squares estimates were found by minimizing the sum, $\displaystyle\sum_{j=t_b}^{j=t_d} \sum_{i=1}^{i=n_j} \left[\tilde{F}_j(\theta_0) - \tilde{f}_i(j) \right]^2$. From this first bootstrap sample, one bootstrap estimate of the parameter vector, θ_i, was obtained. One hundred bootstrap samples were then generated, and their mean was used as the final parameter estimate, $\hat{\theta}$. These least-squares estimates treat the vials as the units of observation. Since the number of vials used was limited at the early ages, these regressions do not weight the very early ages heavily, although the very late ages contribute less to minimizing the squared deviations due to the small number of survivors at those ages.

Our analysis of the EHF models using individual survival and group fecundity data from Rauser et al. (2006b) found that, except for 1 case out of the 10 examined, the four-parameter model (Equation 9-5) had the smallest values of both AIC and BIC (see the Appendix for statistical details of this analysis and Table A9-2). Accordingly, we focused on this model in our detailed analysis of the CO data.

Fecundity and mortality rates were measured for each of the five replicate CO populations. Figure 9-6 shows the data for both age-specific fecundity and female mortality, along with their respective fitted models for all five populations. Although the fecundity model is composed entirely of linear functions, the fact that the population is composed of two types of females, the normal and the dying, produces predicted fecundities that decline slowly and in a non-linear fashion with age (Figure 9-6). The age of onset of the late-life fecundity and mortality-rate plateaus for a population, with their respective breakdays, were estimated from the stochastic fecundity model and the two-stage Gompertz model, respectively.

Group Fecundity Records Only

When only fecundity data from groups of females exist, it isn't possible to estimate all of the parameters in Equation 9-5. However, using the fecundity data alone, we can get estimates of the parameters for Equation 9-3 using

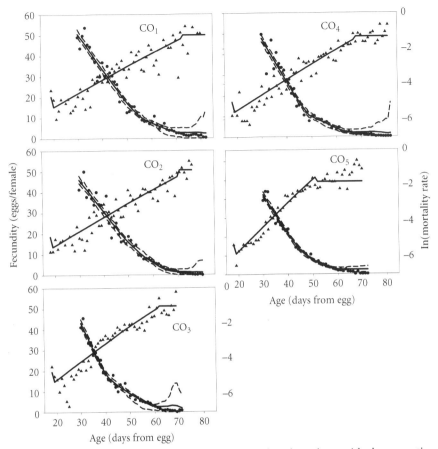

Figure 9-6. Age-specific female mortality and fecundity data along with the respective model predictions for all five CO populations. The circles are the observed mean fecundity and the triangles are the observed mortality rates. A two-stage Gompertz model was fit to the mortality data, and the four-parameter stochastic fecundity model was fit to the fecundity data to determine the breakdays, or the onset of the late-life plateaus, for both mortality and fecundity. The dashed lines are the upper and lower 96% confidence intervals for the fecundity predictions. Fecundity plateaued earlier than mortality in all five populations. The average pairwise difference between the onset of the two types of plateaus was 12.7 days.

standard nonlinear regression techniques. From these we can use the estimated breakdays to make important evolutionary inferences. For this procedure to be valid, it is important to assume that there is some correspondence between the estimated value of the breakday utilizing only Equation 9-3 versus the value for the breakday derived from the full model (Equation 9-5). We explore this problem below.

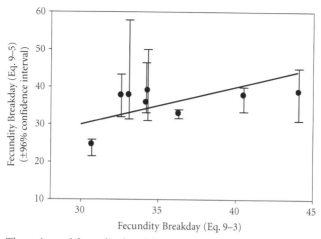

Figure 9-7. The estimated fecundity breakday for eight CO cohorts. For each population, the breakday was estimated by Equation 9-3 (*x*-axis) and Equation 9-5 (*y*-axis). The solid line shows equality of the two estimates.

For eight different experimental data sets, we estimated the parameters of Equation 9-3 from the fecundity data only. Three of these data sets, referred to as CO_{1-1}, CO_{1-2}, and CO_{1-3}, are replicate experiments on individual females from the CO_1 population (Rauser et al. 2005a). For these three cohorts, the parameters of Equation 9-3 were also estimated from the EHF model (Equation 9-5) using the techniques described previously for individual fecundity and survival records. The remaining five cohorts were obtained from the entire set of five CO populations (CO_{1-5}). However, for these data, fecundity was recorded on groups of females and survival was observed on a separate group of females. Accordingly, the parameters of the stochastic fecundity model were estimated by the techniques described previously for individual survival records and group fecundity records. These analyses were done using an adult-age time scale. Thus, time zero is the start of adult life.

The results (Figure 9-7) suggest that the values for the fecundity breakday are very similar with each technique. The other parameters of the fecundity model may have estimates that vary, depending on the technique used. This is not surprising, since the decline in female fecundity with age is described by two parameters in the EHF model (Equation 9-5), while the simple model (Equation 9-5) summarizes this decline with just one parameter. Given these findings, we suggest that reasonable estimates of the breakday for the EHF model can be obtained when survival data are absent by simply fitting the two-stage fecundity model.

EARLY FECUNDITY AND AN ALTERNATIVE MODEL

The EHF model does not treat the changes in *Drosophila* fecundity that take place immediately after sexual maturation. As discussed by Novoseltsev et al. (2003), the initial rapid rise in female fecundity after sexual maturity may represent the balance of ovariole maturation and egg production, with several days being required before females hit their maximum egg output. We believe that the falling force of natural selection acting on fecundity will result in a slow decline in age-specific fecundity after this initial maturation of the female's reproductive physiology. Novoseltsev et al. (2003) suggest that there will in fact be a plateau for some extended period of time at the female's maximum egg production. Novoseltsev et al. (2003) recognize that the pattern of average female fecundity does not typically show such a plateau, but they argue that this is an artifact of the averaging over many females that have plateaus of different lengths.

Novoseltsev et al. (2003) estimate these plateaus by fitting a model to each individual female's age-specific fecundity. An unresolved problem with this approach is whether individual data are sufficiently reliable to distinguish between patterns of fecundity that plateau versus those that show a simple peak with a lot of noise. Future work will hopefully focus on sorting out the different predictions of the resource-allocation model of Novoseltsev et al. (2003) and our evolutionary heterogeneity fecundity (EHF) model.

CONCLUSION: DEATH SPIRALS AND THE EVOLUTION OF LATE-LIFE FECUNDITY PLATEAUS

For the chief subject matter of this book, the most important scientific issue arising from the discovery of death spirals in *Drosophila* is whether or not death spirals invalidate the research that we have performed on fecundity plateaus when we have not had access to complete records of the age-specific fecundities and ages at death of individual females. Recall that this is the case for the fecundity data discussed in Chapters 2, 4, and 5, none of which were based on the collection of complete demographic data for individual females.

We have already used the EHF model to account for death-spiral effects in our fecundity data from previous chapters. Fortunately, the basic evolutionary inferences that we had made before developing the EHF model remain valid; late-life fecundity plateaus evolve in conformity with expectations derived from Hamiltonian models.

10

Physiology of Late Life

Hamiltonian theory suggests that the physiology of late life could be different from that of aging. Drosophila experiments comparing the physiology of aging with the physiology of late life corroborate this suggestion, but much work remains to be done.

INTUITIVE INTIMATIONS CONCERNING LATE-LIFE PHYSIOLOGY

As our discussion to this point should have made clear, there are two chief contending explanations for the late-life phenomenon: lifelong heterogeneity and Hamiltonian evolutionary theory. On the first view, aging is an ineluctable process that never ceases, but in elite subcohorts it proceeds so slowly that age-specific mortality rates roughly plateau at very late ages due to the absence of inferior subcohorts. Intuitively, then, it might be expected that the first view implies that physiologically monitoring cohorts undergoing the transition from aging to late life would reveal no definitive transition, although there might be a gradual deceleration in physiological deterioration as inferior subcohorts progressively die off. Thus, lifelong heterogeneity seemingly implies that late-life physiology should not show distinctive properties compared to the physiology of aging.

On the Hamiltonian view, aging largely comes to an end, with much older individuals no longer afflicted by continuing collapses in the forces of natural selection. For this reason, it has been proposed that there is a distinct "third phase" to life, after the end of the aging phase or stage (e.g., Rose et al. 2006). By contrast to the implications of the lifelong heterogeneity theory, it might be supposed intuitively that the Hamiltonian view necessarily implies that "aging" physiology should stabilize during late life, and the characteristic chronological decline in functional attributes that is such a hallmark of aging must come to an end.

Here we will show that neither of these intuitive expectations is correct, both in principle and in some early data collected from our *Drosophila* experimental system. However, this chapter chiefly serves to open a door to a novel arena for research, and the conclusions that we offer are only preliminary in character.

PARADOXES AND INDETERMINACIES ARISING FROM LIFELONG HETEROGENEITY THEORY

There are four major empirical problems with lifelong heterogeneity theory, as we have discussed. First, there is no direct evidence showing that lifelong heterogeneity in key, delimitable, and measurable robustness characters leads to sufficient differential survival to explain late-life mortality rate plateaus, as we have discussed principally in Chapter 6. Second, there are a range of experiments that give results incompatible with specific lifelong heterogeneity theories, which we have reviewed particularly in Chapter 8. Third, there are experiments that corroborate the alternative Hamiltonian theory for late life (e.g., Rose et al. 2002; Rauser et al. 2006b), particularly experiments involving experimental evolution, which we have reviewed in Chapters 4 and 5. Fourth, there are no experiments or other data that evidently falsify the alternative Hamiltonian theory. For these four reasons, based on data, we have argued against the validity of lifelong heterogeneity theory as an explanation for late-life plateaus in mortality and fecundity.

But as a theory, in and of itself, lifelong heterogeneity has both the strengths and weaknesses of indeterminacy. As an ill-defined and unconstrained theory, it can be modified in innumerable ways. We have taken some pains to show its difficulties in previous chapters when it is formulated explicitly, particularly when it is formulated in terms of lifelong heterogeneity in either background age-independent mortality (the A parameter of the Gompertz equation) or the rate of Gompertzian aging (the α parameter). But there are other conceivable, alternative, ad hoc and post hoc demographic models for mortality based on lifelong heterogeneity, which we haven't subjected to the same level of scrutiny.

We note three particularly slippery features of the full range of lifelong heterogeneity theory from the standpoint of the physiological foundations of its presuppositions: (i) It invokes hypothetical hidden variables such as underlying "robustness" characters of some type, or their converse "frailty" characters. Yet such underlying physiological characters are not operationally defined so that they can be measured physiologically in the literature that invokes them. (ii) Lifelong heterogeneity theory then posits hypothetical interactions between such hidden physiological variables and known demographic variables such as age-specific mortality rates. These linkages can take many conceivable forms, including the production of trade-offs between characters mediated by underlying connections involving the hypothesized physiology. (iii) Lifelong heterogeneity theories have potentially unlimited freedom in the composition of these theory elements, such as arbitrarily varying the subcohort number, arbitrarily varying the number of underlying physiological robustness characters, and so on. Taken together, all these elements of flexibility in lifelong heterogeneity theory provide an expansive playground for the mathematical imagination of a demographer, particularly as it is not reliably tethered to physiological particulars.

Lifelong heterogeneity theory thus seems to meet many of Popper's (e.g., Popper 1959) criteria for unfalsifiability. And thus, we can expect lifelong heterogeneity theory to persist in the gerontological and demographic literatures. An often irrefutable and extremely flexible theory, with hidden variables whose correct identification or measurement can always be disputed, will be very difficult to force out of a scientific arena if it is not rejected on methodological grounds.

Turning to the implications of lifelong heterogeneity theory for the *observable* physiology of late life leads to some of the typical features of an unfalsifiable theory. If there is no particular expectation for the relationship(s) between functional physiological characters and hypothesized lifelong differences in demography, then a transition between aging and late life produced by extreme lifelong heterogeneity can lead to a wide spectrum of trajectories for physiological characters. In explicit simulations of age-specific physiological characters, simulations that assume lifelong heterogeneity for robustness, we can generate a variety of curves for underlying physiological characters. Essentially any pattern can be produced: reversal of functional aging, alternating waves of increase and decrease in functional physiology, continuing physiological deterioration during late life, accelerating deterioration during late life, and so on. An example is shown in Figure 10-1. The lack of structural or parametric constraints on the physiological underpinnings of lifelong heterogeneity theories allows such theories to generate a great variety of possible patterns for the physiological transition from aging to late life. For those who enjoy theory untrammeled by the risk of experimental refutation, lifelong heterogeneity theory will be an attractive way to explain and characterize the physiological features of late life.

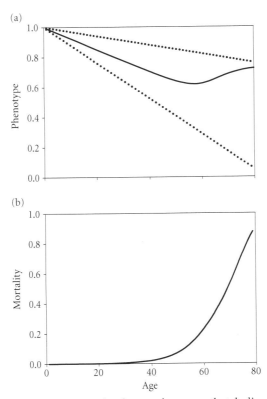

Figure 10-1. A population is assumed to have a phenotype that declines linearly in value with increasing age: dotted lines in (a). There are 101 different physiological types in the population with phenotype-versus-age curves that fall between the two dotted lines in (a). We further assume that as this physiological trait declines with age, so does the age-independent Gompertz parameter for that physiological type. If at age 0 there are equal numbers of all 101 types, then the age-specific mortality in the whole population remains Gompertzian, as shown in (b). The population average phenotype is the solid line in (a). The phenotype will eventually start to decline again at very advanced ages when there is only one physiological type remaining. The shape of this population average curve can be changed under different assumptions about the number of types and the relationship between the physiological phenotype and the Gompertz equation.

POTENTIAL PHYSIOLOGICAL COMPLEXITY OF HAMILTONIAN PHYSIOLOGY DURING LATE LIFE

One of the inherent strengths of the Hamiltonian analysis of aging is that it has always had within it the implicit prediction that any component of age-specific adaptation that is subject to genetic variation is liable to deterioration, thanks to the strongly declining age-specific forces of natural selection shaping both survival and reproduction. In effect, if an adult adaptation is subject to evolutionary genetic age specificity, then it should undergo some degree of deterioration during the phase of adulthood known as *aging*. This deterioration may be delayed relative to the age at which the forces of natural selection start to fall, and it may not be a very pronounced deterioration in some cases, but there should be an overwhelming tendency for physiological deterioration of widely varying kinds to occur.

The only plausible way by which a function can escape this effect would be if there is no evolutionary genetic possibility for age-specific selection. This would be true, for example, of characters that are essentially unaffected by the passage of biological time. Indeed, such characters may underlie the very capacity of old individuals to continue surviving and reproducing, albeit with much higher average rates of age-specific mortality, as in Charlesworth's (2001) analysis of alleles with age-independent beneficial effects.

But turning to the case of late life requires us to leave general expectations behind, except for a negative prediction. If the broad and characteristic physiological declines observed during aging continued unremittingly after the transition to demographic late life, then the kind of Hamiltonian theory that we have proposed here would be implausible. That is because late life is like the period prior to the onset of reproduction, the period of *development*, in that it does not feature consistent changes in the force of natural selection acting on age-specific survival. During development, mortality rates fluctuate but do not show general or consistent trends. During points of vulnerability or transition, such as hatching, birth, molting, first flights of fledglings, and so on, mortality rates may spike upward. But such upward spikes are not sustained, and there are no general mortality or functional patterns that compare with the sustained declines exhibited during aging. Likewise, there are no general expectations concerning specific physiological functions during development. Some capacities may decline as a function of age during development, while others increase. Thus the capacity of the human child to acquire new languages peaks at an early age, perhaps at 3–6 years of age, and then declines, while the capacity of children to learn mathematics generally peaks later. While aging involves general patterns of deterioration across most physiological functions, development does not have such widespread patterns.

Given the existence of plateaus in the forces of natural selection during late life, we cannot make specific predictions about the patterns that age-specific physiological characters will show, except we predict that during late life the consistent physiological deteriorations of aging will no longer prevail. As suggested by Shahrestani et al. (2009), it is possible that functional characters generally stabilize during late life, much as age-specific mortality eventually does. But it is also possible that, as Shahrestani et al. suggested too, physiological characters vary in their trajectories during late life, some continuing to deteriorate, others stabilizing, and perhaps some even improving with age. Broadly speaking, the physiology of late life might be one of general *stabilization* or one of *complexity*. Either is at least conceivable, since Hamiltonian theory only makes clear predictions where components of fitness are concerned, not their underlying physiology.

INITIAL EXPERIMENTS SUGGEST THAT LATE LIFE IS PHYSIOLOGICALLY COMPLEX

Work in the Rose and Mueller *Drosophila* laboratory by P. Shahrestani (Shahrestani et al., in prep.) suggests two things. First, late life is physiologically different from aging. Second, late life is not marked by ubiquitous physiological stabilization. It is instead more akin to the complexity of development.

Shahrestani has provided us with some preliminary results that are as yet unpublished. She followed six populations with well-defined demographic transitions between aging and late life, the IV and B populations of Rose (1984b), also described in some detail in Chapter 4 (see Figure 4-2). These populations were characterized throughout adult life, during both aging and late life, for a variety of functional physiological characters that decline during the aging period. In one of her data analyses, Shahrestani compared the functional trajectories of these characters between aging and late-life phases, as shown in Figure 10-2 for the character of time spent in motion during a 2-minute interval. As can be seen from the figure, this character appears to stabilize during late life. Results like this from Shahrestani's study corroborate the Hamiltonian expectation that the physiology of late life should be different from that of aging. This particular result leaves open the possibility that late life is marked by a general stabilization of functional physiology.

Shahrestani (Shahrestani et al., in prep.) also studied other *Drosophila* characters in both aging and late-life phases. Among these other characters was negative geotaxis, as measured by the percentage of a group of flies that climb up the side of a vial in a finite period of time. As shown in Figure 10-3, there

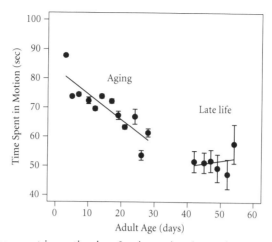

Figure 10-2. Time spent in motion in a 2-minute time interval, measured in seconds, is plotted against adult age. Each point represents the average data from all flies (male and female) from all populations (B_{1-5} and IV) tested at that age. The error bars are standard error of the mean between the six populations. Before age 30, the data points are in the aging phase, as determined by demographic characterization of other individuals in this cohort; after age 40, the points are made up of data collected during the late-life phase. Time spent in motion declines during the aging phase but plateaus in the late-life phase.

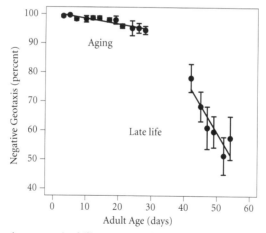

Figure 10-3. Negative geotactic ability, measured as the percentage of flies that made it to the top of an 8-cm glass vial in 1 minute, is plotted against adult age. Each point represents the average data from all flies (male and female) from all populations (B_{1-5} and IV) tested at that age. The error bars are standard error of the mean between the six populations. Before age 30, the data points are in the aging phase, as determined by demographic characterization of other individuals in this cohort; after age 40, the points represent the late-life phase. Negative geotactic ability declined more rapidly in the late-life phase compared to the aging phase.

was no stabilization of this character during the late-life period. Instead, negative geotaxis appears to continue falling during late life, moreover doing so at a faster rate. Again, the pattern during aging was clearly different from that during late life, but the late-life pattern was not one of stabilization. From this, Shahrestani et al. (in prep.) conclude that late life does not constitute a general physiological stabilization. Instead, it can be physiologically complex, much as development is.

ADDITIONAL EXPERIMENTS ON THE HAMILTONIAN EVOLUTIONARY PHYSIOLOGY OF LATE LIFE

From a strictly logical point of view, the two types of data just supplied are key to the physiological interpretation of late life. It appears that neither character shows a continuation of aging-related deterioration, nor is there a general stabilization. But more studies of this kind are needed, with both *Drosophila* and other model organism populations, in order to ascertain the features of late-life physiology more widely. Humans, again, are probably a very poor system with which to address the issue of the physiological transitions from aging to late life, due to differential patterns of medical care and other lifestyle factors that can change as a function of chronological age independently of underlying physiology. But there is no reason why, in principle, the kind of study that Shahrestani has performed could not also be performed on medflies or nematodes, not to mention more arcane model species.

Another type of experiment naturally suggests itself to experimental evolutionists like ourselves. Rose and colleagues have constructed an array of populations with very different ages of transition between aging and late life (e.g., Rose et al. 2002). On Hamiltonian principles, some of the physiological characters that underlie age-specific demographic characters should have undergone corresponding shifts in their transition ages from their *aging trajectory* to their *late-life trajectory*. To use the preliminary Shahrestani results plotted above as a point of reference, populations that have undergone a shift in the average age at which the demographic transition from aging to late life occurs should show at least some corresponding shifts in the ages at which physiological transitions occur. Thus, for time spent in motion, shown in Figure 10-2, the age at which the rapid aging decline of this character effectively plateaus should occur at relatively later ages in replicate populations that have evolved much later mortality-rate plateaus. For negative geotaxis, shown in Figure 10-3, the age at which this character starts to decline more rapidly should undergo a parallel shift when comparing populations that have evolved different starting ages for their late-life mortality rate plateaus. It might be the case that not all such physiological characters will respond at comparable speeds and to comparable magnitudes in the experimental evolution of the age at which aging stops demographically, but at least some cases like this should be significant enough to be detectable experimentally.

Shahrestani has collected data of this kind from populations that have evolved somewhat different starting ages for their late-life plateaus: the CO and ACO populations of Rose et al. (2002). But the extensive data have not yet been analyzed.

PROVISIONAL CONCLUSION: THE PHYSIOLOGY OF LATE LIFE IS HAMILTONIAN BUT COMPLEX

Limited data are available concerning the physiological transition from aging to late life. What we have so far has yet to be published in reviewed journals and must be taken as only preliminary. However, it does suggest that the physiology of late life is broadly Hamiltonian. Functional aging does not merely continue on as if late life is not underlain by different evolutionary rules. On the other hand, Shahrestani et al. (in prep.) have apparently already found enough physiological complexity during late life to suggest that it is as functionally complex as development. Much interesting work remains to be done, and we invite our readers to proceed with it.

11

Late Life in Human Populations

Late life was first detected in human populations, despite the very late occurrence of late life in humans. Recent data from supercentenarians provide evidence for a late-life mortality-rate plateau in human populations. An important evolutionary puzzle is why human populations reach late life so late. Several explanations are conceivable and are not necessarily incompatible with each other. One of these is a generally increased mortality level under evolutionarily novel conditions due to a lack of time for age-independent beneficial substitutions to increase in frequency. Another is a recent expansion in effective population sizes, greatly prolonging the age range over which the effective force of natural selection declines. Regardless of its evolutionary explanation, the cessation of aging in human populations suggests new possibilities for the extension of the human healthspan.

THE PROBLEMATIC NATURE OF HUMAN DATA

We have already mentioned in Chapter 1 that the demography of humans late in adult life has been a common subject of study. It is a commonplace of such studies to note that Gompertzian models start to break down, in terms of their quantitative accuracy, at very late human ages (e.g., Greenwood and Irwin 1939; Gavrilov and Gavrilova 1991). We will be considering the case of the human species in more detail here from the standpoint of the evolutionary biology of late life.

But before we do so, it is important to understand what is and what is not appropriate in discussing the human case. Starting with what is not appropriate is the most important concern in this instance. Humans just are not appropriate experimental animals for obvious ethical and practical reasons. No one should approach the manipulation of human patterns of survival and reproduction with anything but the greatest care and the greatest scruples. Precisely the kind of environmental control and standardization that makes work with organisms like laboratory *Drosophila* and laboratory mice so useful should not be attempted in studies of human biology. Indeed, human biology is being continually shaped by economic, medical, and public health progress, which has measurably improved standards of living, increased life expectancies, and reduced basic human suffering from generation to generation since the eighteenth century. People in seventeenth-century Europe died in great plagues, suffered from unsanitary water, and had no recourse to antiseptic surgery. By every reasonable measure, the lives of Europeans and North Americans have steadily improved since then. And with such secular historical improvements have come rapid and extensive demographic changes.

Madame Jeanne Calment, for example, was born in 1875. For her birth cohort, tuberculosis was still a major risk factor impinging on life expectancy, as were septicemia and any number of other incidental types of bacterial infection. Yet she survived, through two world wars conducted partly in her native France, through the pandemic Spanish influenza that followed World War I, through the invention and widespread adoption of antibiotics, and even through the AIDS pandemic. During that time, even the availability of food fluctuated, particularly with the privations and dislocations associated with the two world wars and the Great Depression. Thus, Calment's individual life history over the course of 122 years occurred in a context of extensive environmental change. Those people who lived for more than 30 or 40 years, in the course of the last two centuries, underwent substantial changes in the environmental hazards that they faced during their lifetimes.

This scientifically inconvenient fact makes human demographic data extremely unsatisfactory from the standpoint of testing fundamental ideas concerning the evolution of life histories, including both the evolution of aging and the evolution of late life. For this reason, our view is that human data should *not* be used for the purpose of making strong inference (Platt 1966) tests concerning the type of scientific theory that we discuss in this book.

On the other hand, the caveats just adduced do not mean that general findings concerning the evolutionary biology of late life cannot be *applied* to human demography. What these caveats mean is that, from the standpoint of scientific inference, there is a one-way street between the type of focused, well-controlled, and theoretically motivated scientific research that is our principal concern, in the first instance, and the type of scattered, uncontrolled, and ad hoc data that are supplied by human demography, in the second instance.

There is nothing special about this stricture. Chemists are not fond of doing experiments outdoors during rainstorms, but that doesn't mean that they can't apply their understanding of chemistry to explain or interpret data concerning acid rain. We regard the application of our research findings to the case of human late life in the same way. We don't regard human data as a useful way to evaluate alternative theories, whether Vaupelian or Hamiltonian, of late life. But we do regard the *application* of Hamiltonian research on late life to the interpretation of human demographic data as a legitimate enterprise.

DO HUMAN DEMOGRAPHIC DATA SHOW SIGNS OF LATE-LIFE MORTALITY-RATE PLATEAUS?

It can be conceded that human demographic data are inherently deficient, but one can still want to know whether, despite that, signs of late-life plateaus can still be detected. This doesn't mean that the failure to detect such plateaus should be regarded as a threat to the scientific salience of late life as a life-historical phenomenon. There is a strict asymmetry here. If human data, with all their deficiencies, nonetheless still show evidence for late-life mortality plateaus, that would suggest the obduracy of the phenomenon, its potential to penetrate the morass of confounds and obscurities by which human demographic data are necessarily afflicted.

Greenwood and Irwin (1939) supplied one of the earliest and most detailed studies of human demographic data from very late in life, working primarily with English actuarial data. Looking chiefly at the mortality records of individuals over 90 years of age, they were led to consider the possibility that "with advancing age the rate of mortality asymptotes to a finite value" (p14) (Figure 11-1). They then approached their human mortality rate data in light of this hypothesis and found that the quantitative fit of late-life actuarial data to this hypothesis is at least reasonable. In particular, they proposed, as a bare possibility, that the rate of human mortality approaches a value of about 50% per year.

It is also notable that Greenwood and Irwin (1939) proposed a crude version of the lifelong heterogeneity hypothesis, invoking Pearl's earlier work on the

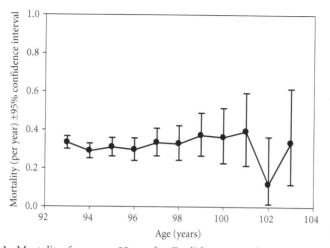

Figure 11-1. Mortality from age 93 on for English women from 1900 to 1920. The mortality data are plotted only for those ages with at least 10 females at risk (from Greenwood and Irwin 1939).

demography of laboratory *Drosophila* mutants, particularly the contrasting demography of wild-type and vestigial mutant flies. They offered, as a thought experiment, a scenario in which a cohort of fruit flies consists of a mixture of wild-type and vestigial flies, the latter dying off entirely so that, at the end of the composite cohort's period of observation, the mortality pattern is defined entirely by that of the wild-type flies. Under these conditions, they supposed that one would observe mortality rate deceleration without having to suppose that aging stops.

Indeed, they assumed that, "In a labile, highly specialized metazoan, decay must surely continue" (p. 14). Thus, while the data surveyed and modeled by Greenwood and Irwin (1939) seem to show an asymptotic approach to a constant mortality rate, and thus a cessation of aging, their reaction was to assume that some type of complication or artifact was responsible. In particular, like later authors (e.g., Maynard Smith et al. 1999), they proposed that individuals over the age of 90 years are likely to undergo a change in their circumstances and thus enjoy a mitigation in mortality. In these respects, Greenwood and Irwin's analysis anticipates major protective maneuvers that have been used repeatedly over the last 70 years to safeguard the near-universal assumption among biologists, gerontologists, and demographers that aging continues unabated at later ages, the demographic data from very old humans notwithstanding.

It is historically interesting that Comfort (1964, Figure 18) supplied a graphical plotting of one of Greenwood and Irwin's (1939) data tables, showing

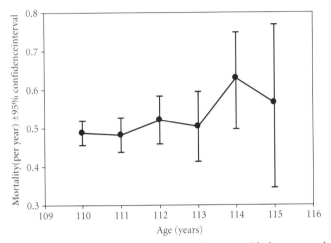

Figure 11-2. Mortality from age 110 on for men and women with documented ages above 110 years as determined by the Gerontology Research Group (data from www.grg.org/ Adams/I.HTM). The mortality rate is plotted only for those ages with at least 10 individuals at risk (Young et al. 2009).

how well the numbers fit a simple exponential decay pattern, a pattern that implies a lack of demographic aging after the age of 90 or so. Gavrilov and Gavrilova (1991) examined much more extensive European demographic data and likewise found that Gompertzian models break down at sufficiently late ages.

Recently, Young et al. (2009) have supplied a graphical compendium of human global mortality rates after the age of 110; an adaptation of their figure is shown here as our Figure 11-2, focusing only on mortality rates up to 115 years of age. After this age, there are so few individuals that the observed mortality rates fluctuate wildly between 0 and 100%.

From this figure, it is evident that Greenwood and Irwin's (1939) suggestion of an asymptotic human mortality rate of about 50% is well within the bounds of statistical plausibility. It should be noted, further, that 984 individual deaths contribute to the pattern shown in Figure 11-2, so this result is likely to remain stable, at least qualitatively, as more deaths are recorded among individuals over the age of 110 years.

Nonetheless, for the reasons already adduced, these data hardly have the scientific quality of laboratory animal cohort studies. In particular, these data are compiled from actuarial records from multiple countries, chiefly European, together with those from the United States, Canada, Australia, and Japan. Though all these countries are affluent, they have significant disparities in health care delivery and access. Therefore, some caution is warranted in viewing these data. Still, Greenwood and Irwin's (1939) purely demographic conjecture seems to be sustained. Age-specific human mortality rates seem to asymptote to an approximately constant value. Thus, demographically, human aging too appears to cease.

As difficult as it is to collect good human actuarial data, the likelihood of collecting good functional or physiological data on the transition from aging to late life in human populations is even smaller. Our position going forward, then, is to accept provisionally the hypothesis that in humans, just as in well-studied fruit flies, aging does in fact cease at the level of the individual, and this cessation leads to the demographic patterns that have been noted for the last 70 years. With this provisional conclusion, we turn to a discussion of its interpretation and its possible practical implications.

WHY DOES HUMAN LATE LIFE BEGIN SO LATE?

It is significant for the history of science that late life begins so late in humans. If it started at age 35 or 40, then it would have become apparent at least as early as the nineteenth century, when good actuarial tables became generally available, that demographic aging can cease. Human aging has long been the intuitive source of much reasoning, and certainly still more emotional concern, about aging in general. As late life is not apparent in human actuarial records until the cohort decade that begins at 90 years of age, the intuitive notion of unceasing aging took hold millennia ago and still grips both popular and scientific imaginations.

But in the demographic patterns of the medflies studied by Carey et al. (1992), as well as those of some of the fruit fly populations studied by both Curtsinger's laboratory (e.g., Curtsinger et al. 1992) and the Rose laboratory (e.g., Rose et al. 2002), late life starts much earlier relative to the duration of development. In humans, development to first reproductive maturity takes about 12 to 14 years. Late life then starts at 90 to 105 years, about eight times later. In our B fruit flies, development to first reproductive maturity takes about 10 days, while the transition to late life occurs at about 38 days from the egg stage of life (Rose et al. 2002). To scale this pattern to human data, this would put the transition to human late life in the decade between 40 and 50 years of age. Thus, human late life begins about twice as late as that of some fruit fly populations, relative to the duration of development to the point of first reproduction.

On the other hand, Rose et al. (2002) show that their O populations, which take only about a day longer to develop than the aforementioned B populations (Chippindale et al. 1994), start late life at the age of about 72 to 82 days from egg. This is a pattern more like that of human populations. In the case of the B and O populations studied by Rose et al. (2002), we have a fairly good idea as to why late life starts at such different ages: the last ages of reproduction in their evolutionary culture histories were different for hundreds of B generations. This raises the question: why have humans ostensibly evolved such late onset of late life?

There are two broad types of evolutionary answers to this question on first inspection: selection history and demographic history. Both of these types of answers must take into account an important feature of the cultural and evolutionary history of *Homo sapiens*: the recent adoption of an agricultural way of life by the majority of the species. It is generally agreed that humans widely adopted agriculture as their primary mode of sustenance on the order of 10,000 years ago (Lindeberg 2010). If we assume an average generation length of about 25 years among human populations, then human populations have undergone selection under agricultural conditions for at most about 400 generations. And many human populations have adopted agricultural life much more recently, in some cases only in the last 5 to 10 generations. The transition to agriculture necessarily generated selection for adaptation to a different diet, a different distribution of mortality risks, very different population densities, and very different patterns of migration between local breeding demes. This is the backdrop to any discussion of the selection and demographic histories that underlie the evolution of human life history.

In this evolutionary and historical context for human aging and late life, there are several alternative, although not mutually exclusive, evolutionary genetic mechanisms that have come into play:

1. Humans have undergone a process of adaptation to the agricultural way of life with respect to the impact of diets based on grasses and dairy products on human health, function, reproduction, and chronic disease. This process of adaptation may or may not be complete as a function of the total time since each human population shifted from a hunter-gatherer to an agricultural way of life.

2. Humans have undergone adaptation to altered demographic patterns of survival and reproduction, with changes to the first age of reproduction, the last age of reproduction, and the shape of Hamilton's forces of natural selection between those two ages. (Note that all three of these features of the forces of natural selection could be different for the two sexes, both before the adoption of agriculture and since.)

3. Humans have undergone evolution in response to large-scale changes in effective population size brought on by agriculture, with the most

likely pattern being a substantial increase in local deme size relative to pre-agricultural population structure.

We will now discuss each of these factors in order, although what follows should be taken as the start of a process of scientific evaluation rather than a summative conclusion.

THE EFFECT OF DIETARY CHANGE ON HUMAN AGING AND LATE LIFE

Lindeberg (2010) has recently contributed an extensive discussion of the relationship between different types of human diets and chronic human diseases, particularly those that could be considered age-associated diseases. A key conclusion of his book is that many features of the age-dependent pathophysiology of chronic human diseases, such as cardiovascular disease or metabolic syndrome, arise from the agricultural diet. Evidently, this mode of reasoning is based on the concept of inadequate adaptation to the agricultural diet, with significant benefits to be achieved by switching back to a diet that resembles that of a hunter-gatherer. Such arguments are founded on an assumption of incomplete adaptation to agricultural diets.

Research on experimental evolution provides a useful perspective from which to evaluate arguments like those of Lindeberg (2010), which are not unique in the anthropological or epidemiological literature. Focused, sustained, and intense laboratory selection is sufficient to change functional characters rapidly (vid. Garland and Rose 2009). That is, when natural selection is very strong, experimental evolutionists would expect 200–400 generations of selection for adaptation to a particular environment to be sufficient to produce extensive and effective improvements to the level of fitness required to function in a novel selective environment. Data from the laboratories of Matos (e.g., Simões et al. 2009) and Rose (e.g., Rose et al. 2004), in particular, have shown a pattern of very rapid adaptation to novel conditions. This type of result seemingly impinges on the view of Lindeberg (2010) and, for the present purpose, the relevance of the historical transition in human diet to the interpretation of human aging, including the transition from aging to late life.

But there is an important omission in this line of reasoning. Given Hamilton's forces of natural selection, adaptation to a novel environment will scale according to age when there is age specificity to at least some of the genetic variation that underlies such adaptation to the novel environment during adulthood. That is, Hamilton's forces scale the intensity of natural selection such that, qualitatively, we can expect adaptation to agricultural diets to have proceeded very effectively at early ages, such as those ages before and just after the first age of reproduction. But at later adult ages, we should expect to see a quantitative and progressive reduction in the extent of adaptation to the agricultural diet. With this factor in mind, we should expect a failure of Lindeberg's (2010) reasoning at juvenile and early-adult ages, but much greater applicability at later adult ages, for populations that have long sustained themselves agriculturally. A crude but perhaps evocative way to convey this Hamiltonian effect on human health is to say that, as one chronologically ages during adulthood, one is proceeding backward in evolutionary time. As the human body undergoes this form of "evolutionary time travel," the lack of adaptation to the agricultural diet will thus become steadily more important.

This argument is particularly important when considering the transition from aging to late life. In effect, human aging is amplified when it is viewed as the detuning of age-specific adaptation during the course of the first phase of adulthood. There is the basic pattern of aging generated by the Hamiltonian reduction in age-specific adaptation as adult age increases; this must produce numerous accelerating forms of pathophysiology as a function of age. And added to this is a pattern of detuned adaptation to the novel agricultural diet. This combination could have been a factor in producing our much later transition from aging to late life, particularly given the pleiotropic echoes of these two progressive detunings of adaptation acting in conjunction. However, this type of hypothesis needs further study, using both explicit evolutionary genetic simulations and laboratory evolution experiments in which the age-dependent impact of adaptation to a novel environment can be studied explicitly.

THE EFFECT OF AGRICULTURAL LIFE-HISTORY CHANGE ON HAMILTON'S FORCES OF NATURAL SELECTION IN HUMAN POPULATIONS

Regardless of the salience of views like those of Lindeberg (2010) and others (e.g., Eaton and Konner 1985) regarding the effects of qualitative human dietary change in human evolution, the adoption of an agricultural way of life must have radically changed the demographic patterns of human survival and reproduction. One way to explain the very late transition from aging to late life in human populations would be to propose that the demographic effects of agriculture were analogous to the transition from early-life reproduction to later-life reproduction, which has been such a staple of laboratory evolution experiments on aging in *Drosophila* (e.g., Rose et al. 2004). Thus, to give one scenario, it could be argued that the adoption of the agricultural way of life might have led to three consequential changes in human life histories: (i) postponed age of onset of reproduction, (ii) increased rate of survival during adulthood, and (iii) increased fertility of humans, perhaps particularly males, at later adult ages. Together, these effects would have been fully parallel to the life-historical regimes characteristically imposed on populations like the *D. melanogaster* O populations of Rose et al. (2002), populations that show a much later age of transition from aging to late life.

The simulations that we have provided in Chapter 3 (Figures 3-2 and 3-5) or those of Rose et al. (2002) illustrate what happens during such evolutionary transitions as a result of altered life histories: a progressive wave of age-specific adaptation at later and later adult ages that, in turn, postpones the transition from aging to late life. The advantage of this type of explanation is that it does not depend on any feature of the evolutionary process that remains to be worked out. As we have shown in this book, using both numerical simulations and data from experimental evolution, this evolutionary mechanism can readily generate a shift in the age at which aging stops. This isn't to claim that this evolutionary scenario is in fact the best, or the only, scenario for explaining the very late transition from aging to late life in human populations; the present authors are not aware of anthropological data that could confirm this hypothesized transition, as the nature of preagricultural demography itself remains a point of some controversy (see Panter-Brick et al. 2001 for an interdisciplinary review of preagricultural society).

EFFECT OF INCREASED EFFECTIVE POPULATION SIZE ON THE AGE OF TRANSITION TO LATE LIFE

A surprising result of our explicit simulations of the evolution of the transition from aging to late life was that this age depended critically on effective population size. In particular, it was initially counterintuitive that a smaller effective population size produced an earlier transition to late life, as shown in Mueller and Rose (1996, Figure 2). In retrospect, this effect can be understood intuitively: reducing the effective population size reduces the width of the range of adult ages over which selection has a differential impact among ages, bringing forward the first age at which age-specific characters are no longer differentiated with respect to the impact of natural selection. More formally, the strength of selection (β) of new mutants will become smaller with reduced population size. Thus, the frequency of strongly selected mutants (see Figure A3-4) will be reduced. Mutants with small effects are more likely to have their phenotypic effects on fitness restricted at late ages; hence, the range of ages not under the influence of selection increases.

In the context of the transition from hunter-gatherer life to agricultural life, there is little question that this produced broadly progressive increases in human population densities. This effect alone, then, in view of the simulated effects just described, must have increased the "effective" last age of reproduction in human populations. This population-size effect, in turn, is expected to lead to a delay in the transition from aging to late life in human populations.

Thus, we have three possible evolutionary mechanisms that can explain the relatively late transition from aging to late life in contemporary human populations: (i) qualitative dietary and other lifestyle changes; (ii) a demographic shift to later ages of reproduction; and (iii) increased population sizes increasing the range of ages over which natural selection is effective. It is not our goal at this point to decide which of these mechanisms is predominant, but we believe that further research on these alternative mechanisms would be of great interest.

THE PROSPECTS FOR RADICAL HUMAN LIFE EXTENSION

It is a simple demographic point that greatly extending the human functional lifespan, or *healthspan*, would be much more easily achieved by shifting the age at which human aging stops to much earlier ages. If the process of aging were stopped at the age of 40 years, for example, then the capacity of modern medicine to sustain the survival and function of people over that age would be greatly increased. This does not mean that heart attacks, strokes, or cancer would no longer continue to occur in individuals whose aging has been arrested at that age. Rather, such individuals would continue to suffer from accidents impinging on their cardiovascular functioning, such as a wayward thrombotic plaque or the somatic mutation of a few cells in their lymphatic system producing a lymphoma. But the *rate* at which such health "accidents" occur would not continue its exponential rise. This raises the possibility of medical interventions "rescuing" those individuals whose aging has been arrested soon enough that they can be largely or entirely restored to the level of health they had before the onset of a particular cardiac disorder, malignant tumor, or other medical problem. In effect, this would allow indefinite survival, *provided that* the aging process had been arrested at sufficiently early ages relative to the restorative powers of the available medical treatments.

The obvious question that this scenario raises is, *how* could we arrest human aging at an earlier age than those ages at which it decelerates to a stop now, between 90 and 105 years? For an organism that has had a stable evolutionary regime for some time, there is no certain or obvious answer to that question short of using experimental evolution.

But humans present a different, and intriguing, possibility. We have only recently started to adapt to agricultural conditions. And that qualitative dietary transition has been associated with a demographic revolution in our effective population sizes, our population structure, and our forces of natural selection, as already discussed in this chapter. Is it possible that we could shift our aging pattern to one in which the incidence of chronic age-associated diseases would be greatly reduced *if* we adopt a lifestyle more like that of our hunter-gatherer ancestors? Lindeberg (2010), for one, evidently thinks so. From the evidence presented in this book, however, more than our aging might be affected by this lifestyle transition. It is worth at least mentioning that such a lifestyle transition might also change the age at which the process of aging stops, moving that age to an earlier point. If an effect like this were sufficiently great in magnitude, then the stabilization in underlying health achieved by a reversion to our ancestral way of life in most respects could be sufficient to, first, shift the age at which our aging stops to an earlier age; second, reduce our plateau mortality rate from then on; and third, thereby extend the human healthspan to a remarkable degree, following the scenario described in the previous paragraph. This bare possibility rests on several questionable conjectures, or at least conjectures about which some degree of doubt is reasonable.

But the existence of an age at which human aging stops is not a questionable conjecture, nor is its potential malleability. The cessation of aging is neither a mythological possibility nor an unchangeable feature of life history. Like most features of biological diversity, it is a tunable product of evolution. In principle, anything microevolution can readily change can be modified with the application of enough medical technology. This makes the idea of radically extending the human lifespan by changing the age at which human aging stops of potentially great practical significance, even if we do not now know precisely how to achieve that objective.

12

Aging Stops: Late Life, Evolutionary Biology, and Gerontology

Most biologists have assumed that aging proceeds progressively and unrelentingly until all organisms in a cohort are dead. This assumption has given rise to the widespread view that the underlying physiology of aging is one of unremittingly cumulative damage and disharmony. With the demonstration that later adult life commonly does not have such features, the entire field of aging research now must be recast, both with respect to its characteristic physiological hypotheses and with respect to its relationship to evolutionary biology. Aging is the age-specific detuning of adaptation, not a cumulative physiological process.

YES, AGING STOPS

Our position is that the formal theory and the experimental data that have been presented to this point in this book amount to a case for the cessation of aging at the level of individuals, in turn generating the cessation of demographic aging among cohorts. That is, we conclude that there is reasonable support for the hypothesis that individuals who have reached demographic late life have in fact undergone a change in the processes of deterioration that, in their aggregate physiological effect, produce roughly stable average age-specific mortality and fecundity rates for each such individual.

It might be useful if we reduce the fairly convoluted case that we have built to this point to a series of itemized inferences that, together, amount to the gravamen of our brief:

1. In at least some cohorts that are kept under good conditions free of obvious contagious diseases, predators, and environmental extremes, both age-specific mortality and age-specific fecundity can roughly stabilize, on average, at late adult ages.
2. This pattern of stabilization in both age-specific mortality and age-specific fecundity at late ages can be explained theoretically by the eventual plateaus of Hamilton's forces of natural selection, both in principle and in explicit numerical simulations.
3. Critical evolutionary experiments that shift the points at which these plateaus in the forces of natural selection occur lead, in turn, to shifts in the ages at which demographic plateaus start, as predicted by Hamiltonian theory.
4. There is *no* direct empirical evidence that lifelong heterogeneity in robustness generates late-life demographic plateaus in naturally configured cohorts.
5. Evolutionary theory suggests that the massive lifelong heterogeneity required to explain late life on purely demographic grounds is unlikely to exist in natural populations, because natural selection would act to eliminate it.
6. There is empirical evidence that lifelong heterogeneity in robustness does *not* generate late-life demographic plateaus in naturally configured cohorts.
7. Plateaus in late-life fecundity can be obscured by the process of dying.
8. The physiology of late life is complex, but it can be distinctly different from the general pattern of deterioration that characterizes aging.
9. Humans show demographic cessation of aging with respect to late-life age-specific mortality rates, although there are many uncertainties associated with the explanation and interpretation of these data.

Our conclusion is that we have answered our basic question: aging does indeed stop. Thus, aging is followed by a third phase of life, which we call *late life,* a phase that is fundamentally different in its evolutionary foundations and its physiology.

We now turn to the general implications of this conclusion.

PARALLELS BETWEEN THE CESSATION OF AGING
AND THE SPEED OF LIGHT

It is a cliché of the historiography of science to say that Einstein revolutionized physics, supplanting the essentially Newtonian paradigm that had prevailed before 1905, the year Einstein published five groundbreaking papers. Of course, the Newtonian paradigm was initiated by Galileo and greatly improved by Newton's successors, such as LaPlace. But the term *Newtonian* in physics is as fair as the term *Darwinian* is within biology in that it honors the key figure who provided the first well-worked-out foundations for physics so that it could become the highly successful science that it has been over the last few centuries. Similarly, Einstein was not alone in undermining the prevailing Newtonian paradigm. Others, such as Minkowski, supplied better-developed versions of relativistic mechanics. And Einstein was not involved in the key experiments, such as those of Eddington, that supplied the strong inference tests of Einstein's ideas. But the "headline" characterization of the revolutionary effect of Einstein's work is essentially correct.

What is less noticed in the headline version of Einstein's revolution is that it started from empirical paradoxes that most physicists, before 1905, were generally ignoring. Among the most important of these paradoxes was that the speed of light is always *c*, never more or less, even if the object emitting light is not stationary. That is, one of the key assumptions of classical Newtonian mechanics, the additivity of velocities, was clearly violated by photons. Physicists knew this, but they weren't doing anything about it, at least at the level of reexamining their fundamental assumptions about mechanics, energy, space, and time. From this unraveling thread, Einstein disassembled the tapestry of Newtonian mechanics, replacing it with his relativistic mechanics.

Our view, as we have argued before (Rose et al. 2006), is that the cessation of aging is as significant for the fields of gerontology and demography as the constancy of the speed of light is for physics. In this volume, we have mounted a case for the view that aging does indeed stop. Furthermore, we contend that we have reasonably falsified the attempts of those who do not accept our conclusion, and instead believe that the aggregate demographic cessation arises from life-long heterogeneity producing a within-cohort shift toward predominance of individuals who are more robust throughout adult life. However, we concede that there is no limit to the range of novel demographic models that can be constructed to evade our attempts to falsify such lifelong heterogeneity theories. But until new arguments are mounted that reinstate the traditional view of unremitting aging, arguments that we cannot summarily dismiss on substantive or methodological grounds, we assume henceforth that conventional views of aging are as undermined as Newtonian mechanics, with the cessation of aging playing the role of the constancy of the speed of light in undermining the long-standing traditional view of unremitting aging (cf. Rose et al. 2006; Rose 2007).

In this last chapter, we spell out what we see as the extensive consequences of this revolutionary situation.

SIGNIFICANCE FOR EVOLUTIONARY BIOLOGY

In works like *Evolutionary Biology of Aging* (Rose 1991), evolutionary biologists were content to claim to have explained the commonplace observations of gerontology: the ubiquity of aging among strictly ovigerous species, the absence of aging among symmetrically fissile species, the phylogenetic diversity of demographic patterns and physiological mechanisms of aging, and so on. Up to 1991, the intuition of evolutionists was that once Hamilton's forces of natural selection had reached zero, death and sterility were inevitable and likely to occur very soon after that point. But the account of lengthy demographic plateaus in medflies published by Carey et al. in 1992 was a direct assault on that simpleminded interpretation. An interesting historical note is that the appearance of plateaus in demographic data was in fact described well before 1992 (Greenwood and Irwin 1939; Comfort 1964) but was largely ignored, probably because those data came primarily from human populations.

After some discomfiture in the face of the Carey et al. (1992) and Curtsinger et al. (1992) publications, Mueller and Rose (1996) and Charlesworth (2001) were able to right the overturned boat of Hamiltonian theory, as we have explained in this book in some detail, as well as developing Hamiltonian theory further. Hamilton's equations had always allowed the possibility of the evolution of demographic plateaus; evolutionary biologists just hadn't done the numerical and mathematical work required to see this. We think that there are several lessons to be learned within evolutionary biology from this turn of events. These lessons revolve around *why* this failure of intuition occurred.

At the most basic level, the failure of evolutionary biologists like ourselves to understand Hamilton's equations properly is characteristic of scientists generally. Whenever we step away from the direct and formal corollaries of our theories, we are essentially guessing. A scientist who knows the formal theory that is proximal to his or her guess may do a better job than a person, whether a scientist or not, who does not know that formal theory. But there is the possibility that the manner in which scientists use their formal theory tends to lead them astray.

In the case of Hamiltonian theory, we think that the intuitions of evolutionists were blinded by two things: (i) simpleminded extrapolation and (ii) simplified population genetics theory. During the period in which the forces of natural selection are steadily falling, at the start of the adult reproductive period, mortality rates are expected to rise consistently, provided that one's analysis is couched in terms of age-specific genetic effects on mortality rates. The same situation holds for the evolution of age-specific fecundities: they too are expected to fall. These theoretical assumptions give rise to an expected pattern of steadily falling age-specific survival and fecundity characters. It is thus intuitively natural to extend these falling curves to age-specific survival probability and fecundity both achieving zero values in a reasonable finite period of time.

The intellectual sleight of hand in this intuitive inference is the implicit dismissal of interconnectedness between ages. Yet decades of work have revealed abundant evidence for pleiotropic interconnection between life-history characters at different ages, both antagonistic pleiotropy and non-antagonistic pleiotropy, or what de Grey (2007) has called *protagonistic pleiotropy*. Both of these patterns of pleiotropy are, in turn, natural consequences of the interconnectedness of the genomic, proteomic, metabolomic, and other networks that have been detected in abundance over the last decade or so. That is to say, there are few things *less* plausible than the idea of strictly age-specific genetic effects, because genetic effects are generally *not* strictly focused. Rather, they radiate through large networks, and these effects will be dispersed generically in physiological time.

In this book, we have presented experimental evidence for both types of pleiotropy. Late life is antagonistically connected to early adult life-history characters. On one hand, once selection for early reproduction is reintroduced in populations that have not been recently selected for early reproduction, like the reverse-selected derivatives of the O populations we discussed in detail in Chapters 4 and 5, both late-life age-specific mortality and fecundity characters rapidly evolve (Rose et al. 2002; Rauser et al. 2006b). And on the other hand, the strictly positive values of average age-specific mortalities and fecundities during late life are evidence for protagonistic pleiotropy, that is, beneficial effects at later ages arising from selection acting at earlier ages. Otherwise, age-specific survival and fecundity would fall to zero at late ages.

This extensive pleiotropy, in turn, produced a situation in which the non-pleiotropic age-specific intuitions of evolutionary biologists were systematically misleading. Explicit calculations, such as those of Mueller and Rose (1996) or

Charlesworth (2001), were required for evolutionary biologists to abandon their misleading intuitions.

Furthermore, our own initial intuition was that the transition from aging to late life might be an obdurate feature of life history. This intuition was first undermined somewhat by the obvious and extensive differentiation of break-days among our long-standing *Drosophila* populations, as described in Chapter 4. Then this notion was definitively demolished by the speed with which reverse evolution of only 20 or so generations shifted the breakdays of our populations with respect to both mortality (Rose et al. 2002) and fecundity (Rauser et al. 2006b), as described in Chapter 5.

We detect two basic sources for the errors of interpretation that we committed. The first type of error is common among scientists who work in fields with mathematical theories. In order to derive analytical and general results, theoreticians in fields like physics, economics, and population genetics characteristically assume away potential complexities. They are forced to do so because otherwise they face a proliferation of higher-order terms in their equations that not only make the calculation of equilibria and trajectories difficult, they make the evaluation of local and, still worse, global convergence to attractors (which may be either equilibria or trajectories) still harder, at least for the human mind. Such simplification is not necessarily dangerous if the scientists who use these simplified theories realize that there are hazards that arise from such simplification. In particular, if numerical examples are generated computationally that explore the sensitivity of theory outcomes to structural variation in the equations, then scientists may have a fairly good sense of how well they can generalize from their simplified analytical theories.

Problems can, and in the present case demonstrably did, arise when such numerical calculations are not performed. In effect, what the Mueller and Rose (1996) study did was to calculate explicitly what age-structured population genetics theory actually implies when the full life history is considered, going well beyond the period when the forces of natural selection are falling rapidly with respect to age. That is, we moved beyond the "local," or low-adult-age, cases to look at the more global pattern of the evolution of demography, as we present in some detail in Chapter 3. Doing so naturally generated Gompertzian mortality dynamics for the initial period of adulthood, with "bending," or decelerating mortality dynamics, thereafter. This broke down our natural, but simplified and extrapolationist, intuitions with respect to the long-term trajectories of mortality as a function of age. It then led us to perform similar calculations for the evolution of age-specific fecundity (Rauser et al. 2006b, with similar results.

Having obtained these results theoretically and then corroborated them experimentally, we have reached the point where we have still less trust in the hand-waving sort of extrapolations that evolutionists too often make based on highly simplified formal theory. This is not an argument in favor of merely verbal theorizing. We regard that as still *less* reliable than generalization from simplified mathematical theory. Instead, it indicates the need for still more use of computational tools, whether computer-generated algebra or simple numerical iteration of dynamical equations, in the interpretation of formal theories. In a sense, this suggests that evolutionary theory has reached a point like that of modern physics, in which the human mind can no longer generate predictions for experimental results, not even with the aid of analytical theory. Now we have to use explicit computation to lead us to our experimental predictions, not simple verbal formulations.

A general expectation about the evolution of late life was that there ought to be a wall of mortality where in fact survival goes to zero (vid. Pletcher and Curtsinger 1998). This expectation follows from the simple logic that if the fitness effects of changing survival at very late ages are essentially zero, then evolution by natural selection is free to allow these survival rates to decay to zero. However, as outlined in some detail in Chapter 3, the dynamics of these systems do not follow these simple expectations. When mutants have pleiotropic effects on a range of age classes and random genetic drift is taken into account, mortality plateaus persist. In fact, populations will drift away from the evolutionary "optimum" (see Figure A3-5).

GERONTOLOGY BASED ON CUMULATIVE DAMAGE OR PROGRAMMED AGING IS DEFUNCT

We have devoted a great deal of this book to the difficulties facing the major alternative theory that late life arises from lifelong heterogeneity, because we take that theory as the "last stand" of the ubiquitous assumption in gerontology that aging is a cumulative and unremitting process. Our view is that this conventional assumption is erroneous, and that the existence of a late-life phase in which individual organisms themselves undergo stabilization in their capacity to survive and reproduce is a fatal refutation of this theory.

The corollaries of this conclusion, if it is accepted, are extensive and profound. Conventional biochemical, molecular, and cellular theories of aging that presume some ineluctable process of breakdown, akin to rust or progressively increasing disorganization, are, in our opinion, defunct. This does *not* imply the absence of cumulative damage or disrepair. The declining forces of natural selection can lead to failures of repair as part of the deterioration of age-specific adaptation that arises with the falling forces of natural selection. Or they may not. In particular, the mere existence of a possible source of damage, or the demonstration that a particular form of damage or disrepair arises in a particular organism during a particular part of its life history, is not a warrant for inferring that this type of damage will be ubiquitous and continuing among all organisms. Thus, free-radical damage, which undoubtedly does occur in some organisms at some points in their life histories, is not correctly generalized as a universal and ineluctable type of damage that ensures or determines aging. It is merely a biochemical process that conditionally may, or may not, be part of the pattern of aging in a particular species for a particular period. And this reasoning can be extended ad infinitum. The foundations of aging are not to be found in physics or chemistry, but in the patterns of the forces of natural selection.

There is an alternative conventional, indeed long-standing, theory of gerontology in its classic twentieth-century form: the programmed theory of aging. According to this theory, in its original form, species have specific genetic programs that have evolved to cause aging in a predictable manner. This theory is true, in some sense, for the senescence of erythrocytes, flowers, and perhaps worker bees. That is, evolution can select for the deterioration or elimination of specific structures or individual organisms that are part of a larger evolutionary unit, be it a mammalian body, a plant, or a social insect colony. The notion that such programmed deterioration leading to death is a valid general explanation of aging has been rejected by virtually all evolutionary biologists and, to a lesser extent, by many gerontologists. Only when there is a strong type of group integration, such as that among the cells of a multicellular animal, is programmed aging likely to underlie the aging of a particular component cell, structure, or organism.

There is a more recent insinuation that is common in the gerontological literature of today, which has some affinities with the programmed theory of aging. This insinuation is that aging is *regulated*. If this merely means that aging is affected by signaling pathways, then it is a relatively innocuous, if distracting, usage. Adaptation generally involves signaling pathways; thus, its age-specific breakdown among adults as a function of age, meaning aging itself, may involve signaling pathways, just as aging can involve any aspect of adaptation, so long as that aspect of adaptation has some age dependence during adulthood.

Thus, to refer to the "regulation of aging" is *either* a merely verbal flourish, of some comfort to those cell biologists who like to think in terms of the regulation of any biological process, *or* it is an illegitimate attempt to resurrect the programmed theory of aging, when such aging does not refer to the selectively favored deterioration of a component part of a larger group or other type of supervenient unit undergoing natural selection. In either case, the usage is of little substantive value.

The demise of both cogent damage and programmed theories of aging leaves the field of gerontology with little in the way of useful theory for aging outside of the Hamiltonian theory for aging that evolutionary biology can supply. Furthermore, and as a natural consequence of this situation, conducting gerontological experiments without the guidance of evolutionary biology is necessarily hazardous. On the Hamiltonian view articulated throughout this book, the study of aging is the study of the transient breakdown of age-specific adaptation, arising from declining forces of natural selection, a breakdown that can come to an end before the last member of an artificially protected cohort dies. As such, studying the aging phenomenon, including its rate and its cessation, without attention to the methods, strictures, and potential artifacts developed or discovered by evolutionary biologists over the last 150 years risks committing errors both systematic and incidental. And not least among these methodological issues are those that have arisen from Hamiltonian research within evolutionary biology, the research that is squarely founded on the consequences of Hamilton's forces of natural selection (vid. Rose et al. 2007).

Put another way, we believe that gerontology needs to be refounded on Hamiltonian principles and discoveries. This refounding does not imply that well-founded empirical discoveries that are now part of gerontology should be discarded. But the conceptual framework and experimental plans of gerontology should be appropriately recast in Hamiltonian terms.

There is little in the historical, sociological, or psychological study of scientists that suggests that most gerontologists will find the perspective offered here the least bit congenial. Generally, their training is in biochemistry, molecular biology, or cell biology, because the common view among gerontologists since the 1960s is that these three fields supply the foundations for gerontology. In particular, the apparent non-Darwinian biases of the funding agencies that support a large fraction of the gerontological research in biology will not be easily overcome. In any case, the present bias permeating the American biomedical establishment is that the foundations of medicine are to be found in cell and molecular biology.

It often takes a long while for the academic and scientific communities to shift from one prevailing way of thinking about a particular subject to another, even when all of the empirical evidence points in the other direction, as is the case in gerontology. And although good science does not always prevail immediately in the world of academic research, it eventually does. (The cliché is that it does so "one funeral at a time.") While the futilities of mainstream gerontology continue unabated, future generations of all types of biologists will see the errors of its present mainstream as clearly as evolutionary biologists do now.

IMPLICATIONS FOR THE MEDICAL CONTROL
OF HUMAN AGING

Part of the reason that gerontological research without Hamiltonian founda-tions will eventually wither is that it provides few useful leads concerning the medical control of human aging. Indeed, its common presumption that there is anything to be defined specifically as the *aging process* only leads it to method-ological paradoxes and problems. For example, it is often said that "Rather than attempting to treat heart disease or cancer specifically, if we could just stop the underlying aging process itself then medical progress would be much faster." Or, in the imaginatively articulated program of Aubrey de Grey, all we have to do is reverse seven specific types of cell-molecular damage and we will com-pletely recover our youthful health (e.g., de Grey and Rae 2007). Yet, on a Hamiltonian view, there is no physiological process of aging, only a lack of adaptive information built by natural selection at later ages (vid. Rose 2009). And in particular, this failure of adaptation is expected to produce vastly more different kinds of pathophysiology than could conceivably be remedied by repairing just seven specific types of cumulative damage.

Unlike cell-biological prescriptions, which reflect the zombie-bank theories of conventional gerontology, Hamiltonian research on aging has produced the following very promising insights and possibilities for technological interven-tion into the human aging process:

1. As discussed most completely in this book, aging is *not* necessarily an unremitting process that proceeds until all members of a cohort are dead and sterile. This implies that substantial improvements to the physiological machinery that underlies health do not require halting a devastating and accelerating process. Instead of bending down a curve of endlessly accelerating mortality, the control of aging requires instead that we slow processes of deterioration that, in some cases, come to an end on their own during late life.

2. As shown repeatedly in the Hamiltonian research of the last few decades, it is trivially easy for biologists to produce much longer-lived organisms by altering the forces of natural selection. Both the rate of aging and the age at which aging stops can be altered by experimental evolution. And since experimental evolution works through perfectly ordinary changes in allele frequency, there is the prospect of emu-lating the biochemical effects of such allele frequency change by pharmaceutical and other medical interventions. In particular, when genomic tools are available, the utilization of genomic, proteomic, metabolomic, and other types of "omic" information derived from

longer-lived model organisms will provide numerous leads and insights into the best choice of medical intervention, locus by locus, disease by disease, and molecule by molecule. For it is not molecular or cell technologies that are the problem in mainstream gerontology, only the conceptual equipment of that type of gerontology.

ENVOI

While our biomedical colleagues may find our perspective on aging chilling, if not perverse, there is nothing unusual about evolutionary biologists upsetting their colleagues in the rest of biology. One of Darwin's key points in *Origin of Species* was the displacement of the then entirely conventional invocation of theistic special creation of adaptations as a key explanatory tool of biologists. Evidently, Darwin wanted to replace special creation with natural selection. In the same way, we propose to replace the now entirely conventional conceptual edifice of gerontology with one founded on the formal analyses of Hamilton (1966) and Charlesworth (1980, 1994), as well as Mueller and Rose (1996) and Charlesworth (2001), among others. Specifically, we want to replace notions of relentlessly accumulating damage and disharmony with the age-dependent tuning of natural selection by Hamilton's forces of natural selection. The demonstration that aging stops is, for us, the final nail in the coffin of theories that assume that aging is a merely physiological process akin to rust. We invite our gerontological colleagues to join us at the funeral of the twentieth-century version of their field. We think that the twenty-first-century Hamiltonian version will be much more promising, both scientifically and medically. After all, science advances one funeral at a time.

CHAPTER 3

Baudisch's Challenge to Hamiltonian Theory

We need to address a recent challenge to the use of Hamilton's forces of natural selection in evolutionary theory. Specifically, the generality of Hamilton's results has been challenged by Baudisch (2005, 2008). In Hamilton's derivation, the function $s(x)$ was derived by implicitly differentiating the Euler-Lotka equation $[\sum_{y=0}^{\infty} e^{-ry}l(y)m(y)=1]$ and finding the partial derivative $\dfrac{dr}{d\ln p_a}$, where p_a is the survival from age a to age $a+1$. A new and different approach to the problem of aging was suggested by Baudisch (2008, p. 22), who proposed that "Equally reasonable, alternative forms would have been dr/dp_a, dr/dq_a, $dr/d\ln q_a$, or $dr/d\ln \bar{u}_a$," where $q_a = 1 - p_a$, and $\ln \bar{u}_a = -\ln p_a$. Here we address the cogency of this claim that these alternatives are equally reasonable.

   Consider a simple single locus genetic model with two alleles. The three genotypes A_1A_1, A_1A_2, and A_2A_2 differ in their probability of surviving from age a to $a+1$ according to $P_{11}(a)$, $P_{12}(a)$, and $P_{22}(a)$. All other survival probabilities and fecundities are the same. Then in a population nearly fixed for the A_1 allele, the allele frequency dynamics of the rare A_2 allele is approximately

$$\Delta p_2 \cong p_2(1-p_2)\alpha_p s_{11}(a)T_{11}^{-1}, \qquad (A3\text{-}1)$$

where $s_{11}(a)$ is defined in Equation 3-1 using the genotypic survival values for genotype A_1A_1, T_{11} is the generation time produced by the genotypic survival values for genotype A_1A_1, and $\alpha(p)$ is $p_1 \ln P_{12}(a) + p_2 \ln P_{22}(a) - p_1 \ln P_{11}(a) - p_2 \ln P_{12}(a)$. So, the sign of Equation A3-1 is determined by $\alpha(p)$ and the magnitude of the change in p_2 is determined by the product $\alpha(p) \, s_{11}(a)$.

Explicit population genetics models for survival show that the fate of alleles at a single locus is dependent on the genotypic equivalent of $s(x)$ (Charlesworth 1980, pp. 207–208). We do not have equivalent results for the other proposed measures of Baudisch; therefore, her proposed fitness measures do not have equal standing with Hamilton's original measure. Put another way, in terms of explicit population genetics, Baudisch's proposed indices have no well-founded basis.

To make the problem with the theory proposed by Baudisch more concrete, consider the following example. The survival and fertility patterns shown in Figure A3-1a produce very different curves for Hamilton's index of the strength of selection on mortality ($s(x)/T$) and one of Baudisch's indices ($dr/d\ln \bar{u}_a$,

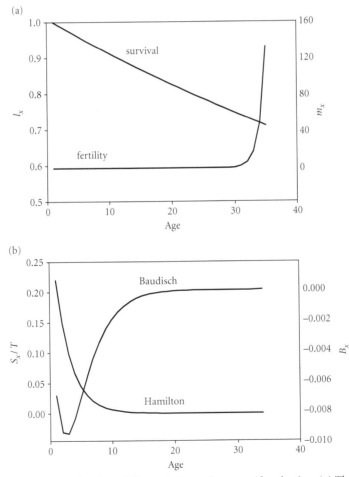

Figure A3-1 An example of two different indices of age-specific selection. (a) The survival (l_x) and fertility (m_x) functions and (b) Hamiliton and Baudisch force of natural selection indices for the life table in (a).

labeled B_x in Figure A3-1b). Hamilton's index shows the expected pattern in the declining strength of selection with age (Figure A3-1b). Baudisch's index shows a maximum impact of selection at age 2, not age 1 (since B_x measures the effects of changing mortality on r it yields negative values, and the larger the negative value, the greater the impact of selection at that age).

To study the behavior of these indices, we followed the change in allele frequencies for 10,000 generations after introducing a mutant with increased survival at age 1, 2, 3, or 4. The expectation is that the greater the force of natural selection, the greater the increase in allele frequency over this fixed period of time. In other words, this analysis does not rely on an index of selection, but rather on the observed dynamics of allele frequency change. We do *not* assume any particular force of natural selection; we let the simulated dynamics determine the outcome of selection. The shape of Hamilton's function in Figure A3-1b, suggests that the allele frequency change should be smaller with each increasing age. Baudisch's index predicts that the strength of selection will be greater at ages 2, 3, and 4 than at age 1.

The outcome of natural selection shown in Figure A3-2 was calculated using Equations 3.14 from Charlesworth (1980), which do not make any assumptions about the strength of selection, unlike Equation A3-1. The outcome of selection follows the qualitative predictions from Hamilton's index and is contrary to Baudisch's expectations (Figure A3-2). We conclude that there is no reason to expect that Baudisch's indices of selection will properly predict the actual effectiveness of natural selection.

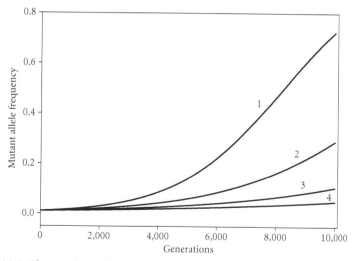

Figure A3-2 The numbered lines show the change in allele frequencies for beneficial mutants that affect age-specific survival at ages 1, 2, 3, or 4, respectively. In each case, the mutant started at a frequency of 0.01.

Numerical Details for Calculations in Figure A3-2

We assumed a single locus with two alleles, A_1 and A_2. The resident population is assumed to be fixed initially for the A_1 allele, and then a small frequency of the A_2 allele is introduced and followed over time. The resident A_1A_1 life history consists of 35 age classes. The maternity function has $m_x = 0.5$ for $x = 1\ldots29$, and $m_x = exp[0.978 \times (30 - x)]$ for all other ages. The survival function is $l_x = 0.99^{x-1}$. The resident population was assumed to be at its stable age distribution.

   The new mutant, A_2, was assumed to affect survival for one age class (p_x). In particular, the heterozygote was assumed to have a survival equal to $1.0025p_x$, while the homozygote mutants had a survival equal to $1.005p_x$. The simulation was started by making the frequency of the A_2 allele 0.01 in all age classes. Allele frequency dynamics were predicted from Equations 3.14 in Charlesworth (1980). We also ran numerical calculations in which mutant survival was changed additively; for example, heterozygote survival was $p_x + 0.0025$ and homozygote survival was $p_x + 0.005$. The results were very similar and are not shown here.

Evolution of Late-Life Simulation Details

The simulations were written in R (version 2.10). Fitness was determined from the single positive real root of Equation 3-2 using the R function *uniroot*. *Uniroot* uses an enhancement of the bisection technique, and therefore does not require that the function be differentiable, but it does require that it have a single maximum in the specified interval (Brent 1973). The solutions returned by *uniroot* had precisions of approximately 7×10^{-9}. This leaves open the possibility that some fitness estimates may differ by less than the precision of our fitness estimates; thus, we would make erroneous conclusions about fitness differences if we used the entire number of machine-significant digits. To circumvent this, we multiplied our solution by 10^8 and converted this real number to an integer, thereby removing all the nonsignificant digits. Of course, mutants whose fitness was the same as that of the resident could only increase in the population by genetic drift.

Relative Roles of Drift and Selection

At each cycle we have three fitness values—resident, heterozygote mutant, homozygote mutant—symbolized as $r_{+/+}$, $r_{+/m}$, and $r_{m/m}$, respectively. We then rescale these fitnesses to 1, $1+sh$, and $1+s$. Although the heterozygote survival is chosen to be exactly intermediate between the resident and mutant homozygote survivals, the fitness is not necessarily intermediate. In fact, we have occasionally found overdominance in fitness. If the initial frequency of the mutant is x_0, then the probability that a new mutant will be fixed is given by

$$\pi(x_0) = \frac{\int_0^{x_0} exp\left[-\beta y\left\{2h + y(1-2h)\right\}\right]dy}{\int_0^1 exp\left[-\beta y\left\{2h + y(1-2h)\right\}\right]dy}, \tag{A3-2}$$

where $\beta = 2N_e s$. In the simulations a uniform random number ($u_i \in (0,1)$) is generated, and if $u_i \leq \pi(x_0)$, then the mutant is fixed by drift (Ewens 1979, Eq. 3.28, p. 83).

We first show the results of a simulation with no genetic drift, so that all beneficial mutants are established, no matter how small their fitness advantage is, and all deleterious and neutral mutants are eliminated (Figure A3-3). Although the results show a plateau in late life with very high mortality, we stopped this simulation after generating 10,000 mutants and we know that we were not at selection equilibrium. The primary point of interest here is the strength of selection on new mutants over the course of evolution.

Let's define the selection coefficient of a new mutant as $s = \frac{r_{m/m}}{r_{+/+}} - 1$. The strength of selection relative to drift is then assessed by the parameter $\beta = 2N_e s$,

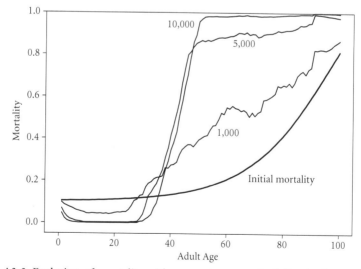

Figure A3-3 Evolution of mortality with no random genetic drift. The lines show the evolved mortality after 1,000, 5,000, and 10,000 cycles of introduced mutants. Mutants were generated according to Equations 3-2 and 3-3 with $\delta = 0.1$ and $\omega = 10$.

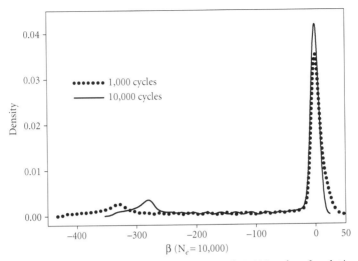

Figure A3-4 The strength of selection (β) after 1,000 and 10,000 cycles of evolution shown in Figure A3-3. β was calculated assuming an effective population size of 10,000 and used the fitness of all possible mutants at each of the two cycles. The selection coefficients were based on the resident phenotype's fitness at each of the two cycles.

where N_e is the effective population size. Neutral genetic variation corresponds to $\beta = 0$, although drift can has a strong effect on the fate of new genetic variation when β is small, i.e. close to zero. The empirical density function of βs after 1,000 introduced mutants and 10,000 introduced mutants is shown in Figure A3-4. After 1,000 introduced mutants, nearly 21% of all possible mutants have $\beta > 1$, whereas after evolution has proceeded through 10,000 introduced mutants, only 1.5% of all possible mutants have $\beta > 1$.

Every 500 cycles in the simulation shown in Figure A3-3, the selection coefficients of all possible mutants was calculated and the largest saved. With these values, the largest values of β are shown in Figure A3-5 over the 10,000 cycles of evolution. These results are consistent with Figure A3-4 in showing that as evolution proceeds, the magnitude of selection weakens. By 4,000 cycles all mutants have βs less than 10, and by 8,000 cycles they are all less than 5 (Figure A3-5).

These results suggest that as evolution proceeds, the impact of drift on the evolution of mortality is increased steadily. Thus, by the time all βs are less than 5, their chance of ultimate fixation is less than 50% even when they start at an initial frequency of 0.1 (Figure A3-6). If the initial frequency of mutants is even less than 0.1, then the role of drift expands over a much greater range of positive mutants (Figure A3-6).

To illustrate the effect of drift, we carried out a new simulation that started from the same initial conditions and the same set of parameters for new mutants as Figure A3-3. However, instead of randomly generating new mutants at each

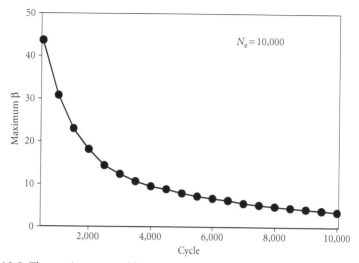

Figure A3-5 The maximum possible strength of selection (β) among all mutants in every 500 cycles of the simulation shown in Figure A3-3. These are not necessarily the actual strength of selection for the mutants that were introduced each cycle. As evolution progresses, it is clear that the fitness range of positive mutants is getting closer to zero.

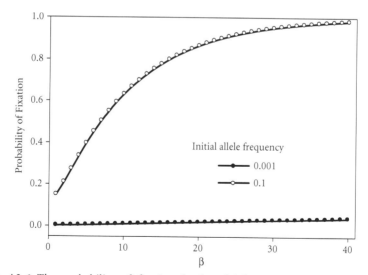

Figure A3-6 The probability of fixation for beneficial mutants at two initial allele frequencies. The probabilities were calculated from Equation A3-3, assuming $h = 0.5$.

cycle, we computed the fitness of every new possible mutant (8,372 total) at each cycle and then chose the mutant with the greatest fitness to be the new resident phenotype. If there was more than one mutant with the maximum fitness, we chose the new resident at random from all the maximum fitness mutants. In this sense, we did a hill-climbing search up to the maximum fitness plateau of this evolutionary process.

The search stopped when the fitness of all new mutants was less than or equal to the resident fitness. The final phenotype reached by this process is show in Figure A3-7 and labeled "Maximum fitness phenotype." At this equilibrium there were 4,102 mutants with the same fitness as the resident. Since the fitness was estimated to eight significant digits, it is possible that there would still be beneficial mutants had we computed fitness more accurately. However, the selection coefficient for these mutants would be on the order of 2×10^{-8}. For β to be greater than 4 and selection to have even a small chance of influencing the fate of these mutants, we would need an effective population size greater than about 100,000,000. This calculation drives home the point that drift will dominate this evolutionary process well before a selection equilibrium is reached.

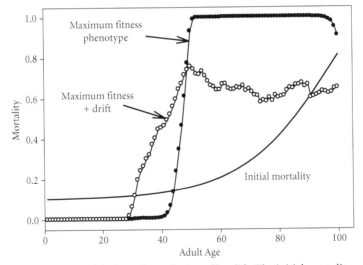

Figure A3-7 The effects of drift on the evolution of late life. The initial mortality schedule was set equal to the schedule shown in Figure A3-3. We then used the maximum fitness searching routine described in the text to find the "Maximum fitness phenotype." The plateau mortality is close to but not equal to 1. We then started a new simulation using the "Maximum fitness phenotype" as the starting condition. We randomly generated mutants and allowed drift to influence their fate. The evolved phenotype from that simulation is labeled "Maximum fitness+drift." The effective population size was 10,000, and the initial frequency of mutants was set to 0.1.

To illustrate the effects of drift, we have taken the "Maximum fitness pheno-type" in Figure A3-7 as the starting point for evolution of random mutants with drift added. The effect is that many neutral mutants and in fact some deleterious mutants now sweep through the population and result in a dramatic decline in the height of the plateau (Figure A3-7, "Maximum fitness+drift" curve).

Other Models of Mortality Plateaus

OPTIMALITY MODELS

Abrams and Ludwig (1995) construct an optimality model based on the dispos-able soma theory of senescence. They take the reproductive schedule of the organism as fixed and assume that the survival schedule is optimized by natural selection. Survival and reproduction are connected through resource allocation. When resources are allocated to reproduction, mortality increases. Thus, if mortality at age i is initially μ_i, after allocation it is $\mu_i + a_i$. Reproduction, $m_i(a)$, is an increasing function of a_i, and $m_i(0) > 0$. Abrams and Ludwig look at a variety of functional forms for $m_i(a)$. Most of the models they consider do not produce plateaus. They interpret late-life plateaus in medflies and fruit flies as consistent with this theory since "The disposable soma theory that we have modeled pre-dicts that aging should cease at an age when reproductive contributions decline to zero.... [T]he leveling of the mortality curve late in life for medflies and fruit-flies...is consistent with aging ceasing after reproduction terminates" (Abrams and Ludwig 1995, p. 1064). However, as shown in this book, reproduction in *Drosophila* does not cease in late life, so optimality theory does not illuminate the cause of plateaus in fruit flies and, as Abrams and Ludwig themselves point out, is apparently contradicted by the very late onset of plateaus in humans.

DIRECTIONALITY THEORY

This theory, developed by Demetrius (1997), is designed for age-structured popula-tions. It assumes that the appropriate measure of fitness is entropy. If $l(x)$ is the prob-ability that an individual survives to age x, $m(x)$ is the mean number of offspring produced by an individual of age x, and r is the intrinsic rate of population increase, then we can define $p(x) = \exp(-rx)\, l(x)\, m(x)$, and entropy ($H$) is defined as

$$\frac{\int_0^\infty p(x)\log p(x)dx}{\int_0^\infty xp(x)dx}.$$

Demetrius (2001) uses an analytical technique also used in the classic paper by Hamilton (1966) to infer the effects of natural selection on survival. He determines the partial derivative of entropy with respect to age-specific survival. Under ecological conditions that limit growth, entropy is expected to increase by natural selection; thus, a positive partial derivative favors the increase of survival, while a negative partial derivative favors a decline.

The results of this theory are illustrated with an example used by Demetrius: human life table data from Sweden in 1835. Figure A3-8 displays the example used by Demetrius (circles) and a slightly altered example with constant mortality in the last three age classes (triangles). We see that even when there is already a mortality plateau, directionality theory predicts that late-life survival

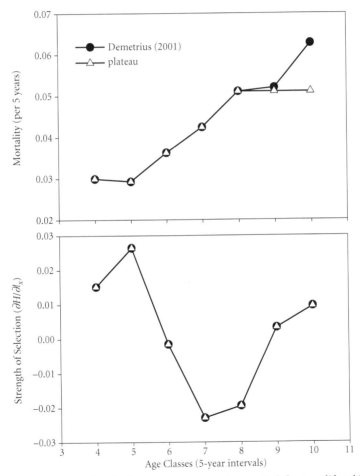

Figure A3-8 The strength of selection on age-specific survival for two life tables. The circles show results for the example given in Table 2 of Demetrius (2001). The triangles are the same data except that the last three age classes have a constant mortality (e.g., are at a plateau). To get the correct mortality for age class 4, the erroneous survival for age class 3 must be changed from 0.94844 to 0.64844 (see Table 2 of Demetrius 2001).

is favored to increase (e.g., the strength of selection is positive in age classes 9 and 10). Increasing survival to age classes 9 and 10 without increasing survival to age classes 7 and 8 can only be accomplished by decreases in late-life mortality. Such decreases would cause an actual drop in late-life mortality.

Although there are occasional unreplicated examples of declines in late-life mortality, this does not seem to be a repeatable or pervasive feature of late-life mortality. For instance, compare the mortality of inbred *Drosophila* lines studied by Fukui et al. (1993) to the populations studied in the same laboratory in a later experiment (Fukui et al. 1996). The former experiment shows that mortality declines at late life, while the latter study only display plateaus with no substantial dips in mortality late in life. Likewise, none of the studies of out-bred populations of *Drosophila* by Rose et al. (2002) show declines in mortality in late life. Thus, this key prediction of directionality theory is not consistent with most empirical data. Under ecological conditions of exponential growth, directionality theory predicts exponential increases in mortality, not plateaus. We conclude that this theory doesn't appear to account adequately for late-life mortality plateaus.

WALL OF MORTALITY

Several commentators have suggested that our simulation results must be flawed since mortality in late life did not reach 100% (Charlesworth and Partridge 1997; Pletcher and Curtsinger 1998). We focus here on one such critique by Pletcher and Curtsinger (1998), since their comments are the most detailed. Pletcher and Curtsinger say that our observation of mortality less than 100% at advanced ages is "inconsistent with the equilibrium predictions of both the antagonistic pleiotropy and mutation accumulation models of senescence, which, under a wide variety of assumptions, predict a 'wall' of mortality rates near 100% at postreproductive ages" (p. 454). But none of the Mueller and Rose models had postreproductive ages. In addition, with respect to the mutation accumulation models developed by Mueller and Rose (1996), in which individuals of all ages reproduced, Mueller and Rose state that "One would also expect that if this process were the only important one determining mortality rates then the mortality rates in the plateau would eventually rise to 100%" (p 15252-15253.).

Nevertheless, Pletcher and Curtsinger make some detailed comments concerning the mutation process and how it affects mortality in our antagonistic pleiotropy models that we will address. Equations 3-2 and 3-3 describe how new mutants affect age-specific survival in our simulations. With these methods, as survival approaches either 0 or 1, the incremental changes to mortality get smaller and smaller. For instance with, $\delta = 0.1$ and $\omega = 10$ when P_x is 0.5, a new beneficial mutant will have a survival value of 0.505. If the initial survival value had been 0.99, the new mutant survival would be only 0.9901.

Now one could argue that, on biological grounds, that it is probably more likely for a new mutant to improve survival by a large amount when survival is low than when it is already very close to 1.

Using our notation, the survival values of the mutants used in Pletcher and Curtsinger were changed to

$$\tilde{P}_x = P_x + \frac{\delta_b}{\omega}, \qquad\qquad\qquad (A3\text{-}3)$$

Deleterious effects were assumed to result in a new age-specific survival value:

$$\tilde{P}_x = P_x - \frac{\delta_d}{\omega}. \qquad\qquad\qquad (A3\text{-}4)$$

Although Pletcher and Curtsinger set $\delta_d = \delta_b = 0.05$ and $\omega = 1$, these equations allow the possibility that the quantitative effect of a beneficial mutation will not be the same as those of a deleterious mutation (e.g., $\delta_d \neq \delta_b$). Equations A3-3 and A3-4 highlight another substantial difference between Pletcher and Curtsinger's work and our own. By assuming that $\omega = 1$, they have eliminated the possibility of mutants affecting multiple age classes. Clearly, an important biological phenomenon that may prevent late-life mortality from reaching a wall of 100% is that mutations affecting these ages also affect earlier ages. In fact, Mueller and Rose (1996, pp. 15252–15253) make this very point by suggesting that a wall of mortality may not occur because "mutations affecting very late survival also have effects early in life, where selection is still effective."

We illustrate the importance of the effect of mutations on multiple age classes with a simple example. Using a simple life history with nine prereproductive age classes and six reproductive age classes, we have simulated the change in adult mortality assuming that mutants have effects across one, two, three, or four consecutive age classes (Figure A3-9), with effects on survival governed by Equations A3-3 and A3-4.

When the window is just a single age class, the first five age classes have mortality reduced to 0 and that of the last age class approaches 1 (Figure A3-9). In these simulations, we did not generate sufficient mutations for mortality to reach 100% in the last age class, but mortality would have evolved to reach that value given enough time. Thus, these results are consistent with Pletcher and Curtsinger's predictions. This result is qualitatively different from those we produce with our models. However, this wall of mortality does not occur when multiple age classes are affected by mutation. When $\omega = 2$, the first four age classes have reached 0 mortality before the last two have hit 1. At this point, no further improvements in fitness are possible, because any reductions in fitness at the last two age classes would have to be accompanied by increases in mortality at earlier age classes or, at best, at the last two age classes, thus exactly

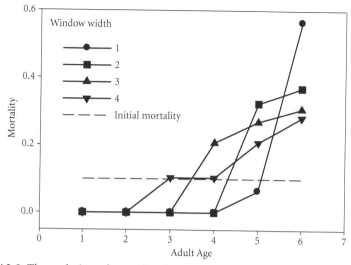

Figure A3-9 The evolution of mortality due to 5,000 mutants with antagonistic effects modeled by Equations A3-3 and A3-4, with $\delta_d = \delta_b = 0.005$. The mutants affected either one, two, three, or four age classes (w).

canceling the benefits. When $\omega = 4$, the simulations converge to a fitness maximum. We determined that evolution had stopped at a local fitness maximum by comparing the fitnesses of all possible new mutants to the resident fitness. The fitnesses of all six alternative mutants are lower than that of the final genotype illustrated in Figure A3-9.

So, even though the Pletcher and Curtsinger conclusions about the wall of mortality are dependent on their special assumption that mutants only affect single age classes, additive models of mutation still present problems for the evolution of plateaus. It is clear that the final mortality pattern in Figure A3-9 is not a mortality plateau.

We next show results for the additive model in which Equations A3-3 and A3-4 determine the properties of new mutants. These simulations assume that mutants affect a large number of adjacent age classes, that is, they have extensive pleiotropic effects. With the additive mutation scheme, it is possible that mortality will hit 100%. When this happens, these simulations truncate the adult life span to the age before mortality hit 100%. Thus, in these evolutionary scenarios, the maximum lifespan may decrease as mutations cause late-life mortality to reach 100%. However, even with the much larger number of age classes in these simulations, it is still relatively simple to compute, at any point in the evolution of these simulated populations, the fitness of every new possible mutant. These fitnesses can then be compared to the fitness of the resident to determine if the evolution is at a local fitness maximum. We have checked the

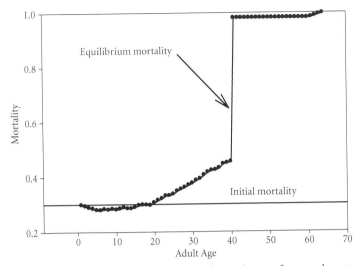

Figure A3-10 Evolution of mortality to a local maximum fitness phenotype. The maximum was reached after 4,000 introduced mutants. The parameters of the simulation were $\delta_d = 0.1 = \delta_b = 0.01$ and $\omega = 60$; fecundity was 5 at all adult ages. There was no genetic drift; the number of initial age classes was 100, and those were reduced to 64 by evolution. At equilibrium the resident fitness was 0.1306774 and the best new mutant fitness was 0.1306627.

fitnesses of all possible new mutants after every 1,000 introduced mutants in order to determine if the current age-specific survival schedule is at a local fitness maximum.

In the example shown in Figure A3-10, the population eventually reached a local fitness maximum characterized by a plateau in late life. While we don't show all of our numerical results for the additive model, we note the following. With small ω, many late age classes evolve to 100% mortality until a maximum fitness is obtained. At these equilibria, early mortality is low and then mortality rises quickly to the maximum age, showing no plateau. With large ω, we also get equilibria with late-age patterns that are largely shaped by initial conditions. Clearly, when initial mortality is relatively constant, late life will evolve plateaus. Finally, none of these results has accounted for the effects of random genetic drift.

In summary, Pletcher and Curtsinger's predicted wall of mortality in late life is seen consistently only if mutations do not have effects on multiple age classes. In addition, their claim that plateaus are contingent on Mueller and Rose's (1996) particular mutation model is not correct. As shown here, the completely additive model of mutation of Pletcher and Curtsinger (Equations A3-3 and A3-4) can result in plateaus (Figure A3-10), albeit under a more limited set of conditions.

Plateaus are transient states

Our primary tool for studying the evolution of late-life mortality has been computer simulations. In our initial paper (Mueller and Rose 1996), we did not check to see if the simulations had reached a loci fitness maximum. Consequently, it has been suggested that the plateaus we observed were in fact simply transient states of a process whose stationary states were not characterized by late-life plateaus. Wachter (1999) focused on the special case of one of our models. He analyzed the antagonistic pleiotropy model, with no genetic drift and mutations that affect a single age class, whereas Mueller and Rose (1996) consider this model and others in which mutations affected multiple ages. Interestingly, Yashin et al. (2000, p. 322) mistakenly inferred that just the opposite was true when they asserted that "Charlesworth and Partridge (1997) and Pletcher and Curtsinger (1998) criticized the assumptions of Mueller and Rose's (1996) models. They argued that it is difficult to imagine any genetic mutation producing changes in the mortality rate at only one or two precise ages. More realistic models of mortality tradeoffs should include changes in survival over larger age intervals."

Wachter (1999) shows that the limiting states of the single age-class model are unlikely to have long stretches of high mortality, only the last few age classes having high mortality rates. In fact, he shows that, if there is a late-life plateau, it is more likely to be with mortality rates close to 0, not 1. Wachter also demonstrates that, for this version of our model, the predicted equilibrium equations derived by Pletcher and Curtsinger (1998, Eq. A5) are wrong.

However, Wachter does not derive results for mutations that affect multiple age classes. Rather, he merely suggests that his results will apply to these models, "Similar arguments are believed to apply to all the Mueller–Rose models" (Wachter 1999, pp. 10546–10547). In practice, what appears to happen is that, long before any equilibrium of the sort Wachter refers to, the progress of selection is halted because the magnitude of the selection on the best possible mutants becomes too small relative to the effects of genetic drift.

Using a model with no drift, dominant mutants, and a window of three age classes, the simulation shown in Figure A3-11 quickly reaches a local maximum in fitness that is characterized by a plateau. In addition, the initial conditions have a linear increase in mortality. Thus, this plateau is not an artifact of our initial conditions. At the calculated equilibrium, all possible mutants have lower fitness than that of the resident genotype except for a genotype that has equal fitness. The one mutant with equal fitness (equal to eight significant digits) results in small in reductions of mortality during the very last two age classes (e.g., from 0.5084763 to 0.5084762 and from 0.5084935 to 0.5084922). This happens because this mutant has both beneficial and deleterious effects hitting the last three age classes, and these effects exactly cancel in age class 4. However, the fitness benefit from this change is less than 10^{-8}. This means that, unless the

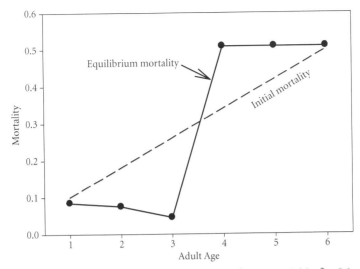

Figure A3-11 Mutations generated by Equations 3-2 and 3-3, $m_x = 1$, $\forall x$, $\delta = 0.1$, and $\omega = 3$. Equilibrium was reached after 821 mutants had been generated. Equilibrium is defined by the condition that the fitness of the final equilibrium mortality phenotype is greater than or equal to the fitness of all possible mutants that can be generated by Equations 3-2 and 3-3 from this equilibrium phenotype.

effective population size is substantially greater than about 10 million, the fate of these mutants would be largely determined by genetic drift. Although it is technically difficult to find a true equilibrium for this model, recall in the previous section that such an equilibrium was found for the completely additive mutation model.

Certainly, with the addition of drift to these models, equilibria of the sort reached in the simulation Figure A3-11 will not be attained. However, it remains for future theoretical work to derive a full analysis of the drift/selection balance stationary distributions.

PLATEAUS ARE A NATURAL CONSEQUENCE OF QUANTITATIVE GENETICS

Fox and Moya-Laraño (2003) suggest that an individual's lifespan can be represented as a quantitative trait determined by the sum of a large number of genes with small additive effects. They then assert that if longevity has a normal distribution, this can be used to derive a new relationship for instantaneous mortality rates that is a function of the mean and distribution of the longevity. The interesting aspect of this model is that observed longevities can be used to estimate the phenotypic mean and variance and therefore the age-specific mortality rates expected in this model. While it is true that the model does give rise to plateaus, they have not identified the mechanism for this condition.

There are at least two major problems that erode our confidence in this model. The first is that the simple assertion that longevity will be determined by small additive effects of many genes does not make it true. Evidently, longevity can be affected by large-effect mutations, as we have already mentioned. Secondly, the data analyzed by Fox and Moya-Laraño (2003) do not generally conform to the model's predictions. The three species whose demography they analyzed show fairly widespread departures from the model predictions, even at very early ages, where there are large numbers of observations and hence reliable mortality estimates. This suggests that their model does not capture important features of age-specific mortality patterns.

EVOLUTIONARY STRING MODELS

Pletcher and Neuhauser (2000) elaborate on a model introduced by Penna (1995) to study the evolution of age-specific mortality. This model shares features of reliability models in that at each age there are believed to be several crucial components that may fail. Failure of all the components at any age leads to death. Genetics are introduced into the haploid model by assuming two genetic states at each locus. The more fit wild-type state has 40 components. A deleterious mutant has only two components. Mutations are allowed in every generation, although each individual gamete may gain only one mutation at one of the 32 age classes per generation. In this respect, this model is a special case of the models considered by Wachter (1999). Simulations were run in which the fates of all individuals in a population of about 3,100 individuals were followed.

This is clearly a novel way to model the biology of an organism, although it should be noted that the model's component assumptions are not readily testable. The model appears to produce plateaus in mortality at late life and to predict that, when a population is subjected to late-life selection (by waiting until age 16 to reproduce rather than age 8), the mortality plateau shifts to later ages.

For this model to be taken seriously, it must be able to reproduce simple patterns of selection that are well established. For instance, since there is no reproduction prior to age 8 in either the early- or late-reproducing populations simulated by Pletcher and Neuhauser (2000), there ought to be equally strong selection for high survival up to age 8. However, Pletcher and Neuhauser's results clearly show that the mortality rates are higher in the early-reproducing population at ages 1–7 compared to the late-reproducing population. Since survival does not affect other fitness components, like fecundity, this result is unexpected. In addition, the equilibrium number of deleterious mutants increases from ages 1 to 8, indicating age specificity in selection prior to reproduction. This result is seemingly contrary to standard population genetic theory. Therefore, it seems difficult to interpret how Pletcher and Neuhauser's evolutionary model is functioning at reproductive ages, which raises questions about its relevance to patterns of mortality in real populations.

MARKOV MORTALITY MODELS

These are models based on Markov processes that are stopped, or killed, at random times. They can be applied to biological longevity by assuming that there is some, possibly unobserved, Markov process that determines an individual's health, the most important features being whether the individual is alive or not. For example, Weitz and Fraser (2001) model what they call *viability* by assuming that it is subject to a constant downward force and random variation with age. Thus, at age $t+1$, an individual's viability is given by

$$v(t+1) = v(t) - \varepsilon + \sigma \chi(t),$$

where ε measures the impact of aging to decrease viability, $\chi(t)$ is the effect of a random process on viability that has a mean of zero and unit variance, and the constant σ modulates the variance of the random process. When viability reaches zero, the individual dies. This type of model can be thought of as a dynamic heterogeneity model, as opposed to the static heterogeneity models we reviewed in Chapter 6. The random effects on longevity are not specifically defined by Weitz and Fraser, but they suggest that these might be phenomena like competition for resources, phenotypic differences, local environmental changes, or even stochastic gene expression. While this list of factors could in principle affect longevity, there is nothing in the formulation of this model that would help us determine how, for instance, increasing or decreasing phenotypic differences will impact mortality. In any case, this very simple model can produce plateaus or even declining mortality at late ages, although, as mentioned before, there is little empirical support for declining mortality with age.

Steinsaltz and Evans (2004) provide a comprehensive overview of these types of models. They refer to these Markov models as part of the evolving heterogeneity theory of mortality tapering. From these theories, Steinsaltz and Evans suggest that these models give us an explanation for mortality plateaus: "As a Markov process progresses, the distribution of its state is being shaped by two forces: random motion spreading the mass out, and shifting it in certain preassigned directions; and deaths lopping off mass at each point, at a fixed rate…if we wait long enough the distribution of those individuals who survive will approach a certain one of these quasistationary distributions. The mortality rate, of course, will also approach the mortality rate averaged over this distribution" (p. 321). When this average mortality rate is reached, these rates will stop increasing with age, and hence we have a plateau.

While this may constitute an explanation of plateaus for a statistician, it is hardly a biological explanation. There have to be biological phenomena that determine the properties of these Markov chains; thus, a biological understanding of this phenomenon is not advanced by a thorough understanding of Markov chain-killing models. Steinsaltz and Evans do make the very useful observation, however, that since there are so many models that can apparently produce plateaus, simply producing a model that can mimic these patterns does not constitute strong support for the model. This viewpoint is very much in line with our own and is, of course, why our research program has focused on experiments that explicitly test our theory in a more stringent manner than merely predicting plateaus that have already been observed.

RELIABILITY THEORIES

These theories are a subset of the Markov chain models discussed above, but they attempt to develop a mechanistic view of an organism, albeit a nongenetic one. The theories of Gavrilov and Gavrilova (2001) are perhaps the most prominent. These models assume that organisms are constructed of many different critical components that have high levels of redundancy. Even if these components themselves don't age, Gavrilov and Gavrilova show that the chance of a critical failure increases with age. They also suggest that mortality rates generated from their models follow the Gompertz pattern. Their models have a number of unrealistic biological assumptions (see Pletcher and Neuhauser 2001), in addition to some technical flaws that invalidate their claim of Gompertz mortality (Steinsaltz and Evans 2004). However, models of this kind might have some use if they are used to model how physiological components of organismal function underlie the evolution of patterns of age-specific survival and fecundity, where the latter processes are supervenient. At this point in time, the only work of this kind of which we are aware is that of Frank (2007) on the development of cancer.

CHAPTER 4

Estimation of Mortality-Rate Plateaus

Mortality rate plateaus were estimated by allowing d^* to be the age at which mortality rates become constant with age (the breakday). Then, at ages x less than d^*, age-specific mortality rates were modeled by the continuous-time Gompertz equation and set equal to $A\exp(\alpha x)$, where A is the age-independent rate of mortality and α is the age-dependent rate of mortality increase. For $x \geq d^*$, mortality rates were assumed to equal $\tilde{A}$. $\tilde{A}$ is independent of age but different from A. For a particular value of d^*, A, α, and $\tilde{A}$ were estimated by maximum likelihood. This was repeated for a range of d^* values, and the value of d^* that yielded the largest likelihood value was chosen as the best estimate of the breakday between early and late mortality.

The likelihood function was constructed from ages at death of the N members of a cohort, following methods similar to those of Mueller et al. (1995). In the experiments of Rose et al. (2002), the raw data consisted of the number of dead flies recorded every 2 days, which might be zero. Therefore, we numbered the 2-day checks sequentially and let t_N be the last check during which the last fly died. Then the number of dead flies in each 2-day period is

$$d_1, d_2, \ldots, d_{t_N}.$$

Likewise, the number of flies alive at the start of each census period is N_1 ($= N$), $N_2, \ldots, N_{t_N}(= d_{t_N})$. Let $q(i)$ be the probability that an individual that lived to census period i dies by census period $i+1$. Then the likelihood function is defined as

$$L = \prod_{i=1}^{i=t_N} \binom{N_i}{d_i} q(i)^{d_i} \left(1 - q(i)\right)^{(N_i - d_i)}.$$

For a particular breakday, d^*, $q(i)$ is then estimated as

$$\begin{cases} 1 - exp\left\{\dfrac{A\left[exp(\alpha 2i) - exp(\alpha 2(i+1))\right]}{\alpha}\right\} & \text{if } 2i < d^* \\ 1 - exp(-2\tilde{A}) & \text{if } 2i \geq d^* \end{cases}.$$

Statistical Tests for the Existence and Evolution of Late-Life Fecundity Plateaus

We tested whether fecundity plateaus at late ages by statistically testing the fit of a model with a late-life plateau to mid- and late-life fecundity data in the cohorts compared in the experiments of Rauser et al. Average population fecundity in *Drosophila* increases at early adult ages until it reaches a peak and then starts to decline. Therefore, we defined midlife as those ages when average population fecundity starts to decline and late life as those ages when the decline in average population fecundity stops or slows. The model we fit to the data was a three-parameter two-stage linear model with the following relationship between age (t) and fecundity ($f(t)$):

$$f(t) = \begin{cases} c_1 + c_2 t & \text{if } t \leq fbd \\ c_1 + c_2\, fbd & \text{if } t > fbd \end{cases}.$$

This model was used since we did not have survival data for all the populations in this study and the techniques for inferring differences in breakdays are easier to apply with this model. A more detailed explanation of this model was presented in Chapter 9.

The model was fitted using all of the fecundity data at each age (100 observations), starting at an age in midlife when the average fecundity for that population started to decline (age 30 days for all CO populations except CO_3, where fecundity did not start to decline until age 46 days, and age 26 days for all ACO populations). Each cohort was fitted to the model independently. This model was fit to the data using a nonlinear least-squares function in the R-project for statistical computing (www.R-project.org). We wrote a self-starting R-function for the two-stage linear model that provided initial estimates for the parameter values as well as the predicted fecundity from the equation.

We tested whether fecundity plateaus evolve according to the last age of survival using the ACO and CO *Drosophila* populations described above. The replicate ACO populations have an earlier age of reproduction and shorter lifespans compared to the CO populations. However, these average lifespan patterns and ages of reproduction by themselves do not indicate the timing or nature of fecundity plateaus for these populations. A pairwise comparison between cohorts obtained from the ACO and CO populations allows us to test properly whether the onset of fecundity plateaus, or the breakday, would occur later in the CO populations relative to the ACO populations.

This experimental design resulted in one ACO_i cohort and one CO_i cohort that were matched by a common index being tested at one time. The common index indicates that the two populations had a common population of origin (O_i). Thus, the pairs of populations form blocks that have a common evolutionary origin and a common set of experimental conditions. Each population also has its own unique history of genetic change due to random genetic drift. Thus, there are three sources of random variation in this experiment: populations, blocks, and individual measurement errors.

In this formulation, index i will indicate one of the 10 populations of origin, j will be one of the five blocks or cohort pairs, and k will be a vial of four individuals, which is the smallest unit of observation within a population. If each cohort has a total of n_i individuals, then the number of eggs per female in population i, block j, individual k is y_{ijk}. The basic nonlinear model is given by

$$y_{ijk} = f(\varphi_{ijk}, v_{ijk}) + \varepsilon_{ijk},$$

where φ_{ijk} is the vector of parameters, v_{ijk} is the covariate vector, and ε_{ijk} is the within-cohort variation. The covariate vector contains the age of individual ijk, t_{ijk}, and the population code, δ_i, which is 0 if the population is ACO (e.g., $i = 1, 2, 3, 4,$ or 5) and 1 if the population is CO (e.g., $i = 6, 7, 8, 9,$ or 10).

For the two-stage linear model, the functional relationship is

$$f(\varphi_{ijk}, v_{ijk}) = \begin{cases} \varphi_{1ij} + \varphi_{2ij}t_{ijk} & \text{if } t_{ijk} \le \varphi_{3ij} \\ \varphi_{1ij} + \varphi_{2ij}\varphi_{3ij} & \text{if } t_{ijk} > \varphi_{3ij} \end{cases}.$$

We assume that the values of the model parameters are affected by both fixed and random effects. The fixed effects can be examined to determine if the selection treatment has a significant effect. The parameters are also assumed to vary randomly between populations due to founder and drift types of effects and between blocks. The between-block variation may be due to either different experimental conditions or founder effects. These two sources of variation cannot be separated. These assumptions translate into the following system of equations:

$$\begin{aligned} \varphi_{1ij} &= \beta_1 + \gamma_1\delta_i + b_{1i} + c_{1j} \\ \varphi_{2ij} &= \beta_2 + \gamma_2\delta_i + b_{2i} + c_{2j} \\ \varphi_{3ij} &= \beta_3 + \gamma_3\delta_i + b_{3i} + c_{3j} \end{aligned} \qquad \text{(A4-1a–c)}$$

where the γ_k ($k = 1$–3) are the fixed effects due to selection, the b_{ki} are the random population effects, and the c_{kj} are the random block effects. An important statistical test will be to determine if the γ_k are significantly different from zero. If so, this will indicate that the selection treatment has a statistically significant effect on the regression model parameter.

Fecundity decreases substantially with age in these populations, which suggests that we should model within-population variance as a function of mean fecundity. The general formulation is

$$\text{Var}(\varepsilon_{ijk}) \cong \sigma^2 g^2(\hat{u}_{ijk}, v_{ijk}, \delta),$$

where $\hat{u}_{ijk} = E(y_{ijk} | \mathbf{b}_i, \mathbf{c}_j)$. In this analysis we used $g(.) = |y_{ikj}|^\delta$, where δ is estimated from the data. The $\mathbf{b}_i$ were assumed to be distributed as

$$\mathbf{b}_i \sim N\left(\mathbf{0}, \begin{bmatrix} \Psi_{11} & 0 & 0 \\ 0 & \Psi_{22} & 0 \\ 0 & 0 & \Psi_{33} \end{bmatrix}\right).$$

The $\mathbf{c}_j$ are assumed to be distributed as

$$c_j \sim N\left(0, \begin{bmatrix} Z_{11} & 0 & 0 \\ 0 & Z_{22} & 0 \\ 0 & 0 & Z_{33} \end{bmatrix}\right)$$

The maximum likelihood techniques used to estimate the model parameters and test their significance are reviewed in Pinheiro and Bates (2000, chapter 7). These techniques were implemented with the nonlinear mixed effects package in R (version 1.6).

Lastly, using the parameter estimates from the model, the height of the late-life fecundity plateau is

$$\hat{\varphi}_4 = \hat{\varphi}_1 + \hat{\varphi}_2\hat{\varphi}_3. \tag{A4-2}$$

Since $\hat{\varphi}_4$ is a nonlinear function of the three estimated parameters, its variance was estimated using the delta method (Mueller and Joshi 2000, p. 83). The variance in plateau height is then

$$\begin{aligned} \text{Var}\left(\varphi_4\right) = \text{Var}\left(\varphi_1\right) + \varphi_3^2\text{Var}\left(\varphi_2\right) + \varphi_2^2\text{Var}\left(\varphi_3\right) + 2\varphi_3\text{Cov}\left(\varphi_1\varphi_2\right) \\ + 2\varphi_2\text{Cov}\left(\varphi_1\varphi_3\right) + 2\varphi_2\varphi_3\text{Cov}\left(\varphi_2\varphi_3\right). \end{aligned} \tag{A4-3}$$

Asymptotic 95% confidence intervals on the plateau height, $\hat{\varphi}_4$, are estimated as $\hat{\varphi}_4 \pm 1.96\sqrt{\text{Var}(\hat{\varphi}_4)}$. The variances and covariances in Equation A4-3 are estimated from the nonlinear least squares procedures.

CHAPTER 6

Methods for the Simulations used in Figure 6-2

A single set of Gompertz parameters were used for all of these simulations, A (0.0725346) and α (0.22891005). Using these Gompertz parameters, 1,000 random ages at death, d_i, were generated using the inverse transform method (Fishman 1996), $d_i = \ln[1 - \alpha\ln(1 - U_i)/A]\alpha$, where $i = 1,\ldots, 1000$ and $U_i \sim$ uniform (0,1). To each d_i, experimental error ε_i was added where $\varepsilon_i \sim N(0,\sigma^2)$. The various values of σ^2 are given in the figure.

Methods for the Simulations Used in Figure 6-3

The mean vector of the Gompertz parameters was $\mu = (A,\alpha)$. We assumed these parameters had a multivariate normal distribution on a natural log scale, with $\Sigma = \text{Cov}[\ln(\mu)]$. We took N samples, X_i, $(i = 1,\ldots, N)$ such that $\ln[X_i] \sim \text{MVN}(\ln(\mu), \Sigma)$.

These N samples constitute one cohort. If $\Sigma = \mathbf{0}$, then the population should obey the Gompertz equation. If $\Sigma \neq \mathbf{0}$, then there is heterogeneity in the population and the possibility of a mortality plateau if the variation is sufficiently large.

For each $\mathbf{X}_i$ we computed a random age at death using the inverse transform, $\ln[1 - \alpha_i \ln(1 - U_i)/A_i]\alpha_i$, where $U_i \sim$ uniform $(0,1)$. From these ages at death we computed 2-day mortality rates. For the simulations with variation in both A and α, we assumed that $\text{Cov}[\ln(A), \ln(\alpha)] = 0$. The parameter values used were $\mu = (0.0725346, 0.22891005)$, $N = 1028$. Σ varied as described in the text. The simulations were programmed in R, version 2.40.

CHAPTER 7

Population Genetic Model

Suppose we have k alleles, $A_1, \ldots, A_k$, with frequencies $x_1, \ldots, x_k$. Each allele is associated with an age-specific survival phenotype, $p_i = (p_1 i, \ldots, p_{di})$, where the total number of adult age classes is d. The relationship between genotype and phenotype is, for homozygotes,

$$A_i A_i \Rightarrow p_i \qquad\qquad (A7\text{-}1)$$

and for heterozygotes it is

$$A_i A_j \Rightarrow p_{ij} \Rightarrow \{p_{sij}\} = max(p_{si}, p_{sj}). \qquad\qquad (A7\text{-}2)$$

In other words, the heterozygotes were assigned the maximum survival exhibited by either of their constituent alleles.

Fitness for genotype $A_i A_j$, w_{ij}, was determined by the largest root of the Lotka equation,

$$\sum_s e^{w_{ij}s} l_{sij} m_s = 1,$$

where l_{sij} is the product of the age-specific survival probabilities. Allele frequency change was then determined by the standard theory. Thus, the frequency of allele i in the next generation, , is given by

$$x_i' = \frac{x_i w_i}{\bar{w}}, \qquad\qquad (A7\text{-}3)$$

where the marginal fitness of allele i, w_i, is defined as $\sum_j x_j w_{ij}$ and the mean fitness, $\bar{w}$, is $\sum_i x_i w_i$.

Computer Simulations

The life history of the simulated organism had 9 preadult age classes and 10 adult age classes. The initial adult mortality was given by the Gompertz equation (Equation 2-1), with $A = 0.01$ and $\alpha = 0.4$. All new mutants affected four adjacent age classes. The first affected age, D, was selected at random. With probability 0.5, age-specific survival of the most common allele's survival phenotype, p_k, would be altered to create the mutant phenotype:

$$p_{ik} + (1 - p_{ik}) \left\{ \frac{c_1 + (c_2 - c_1)}{1 + \exp\left[\dfrac{c_3 - i}{c_4}\right]} \right\}; \qquad \text{(A7-4)}$$

otherwise,

$$p_{ik} \left\{ \frac{c_1 + (c_2 - c_1)}{1 + \exp\left[\dfrac{c_3 - i}{c_4}\right]} \right\}, \qquad \text{(A7-5)}$$

where $c_3 = D + 2$, $c_4 = -2$, and $i = D, D + 1, \ldots, D + 3$. If the correlation in changes was positive, then $c_1 = 0.1$ and $c_2 = 0$; otherwise, $c_1 = 0.05$ and $c_2 = -0.025$. Equations A7-4 and A7-5 were used to generate Figures 7-1 and 7-2. The initial frequency of these mutants was set to 10^{-6}.

After each mutant was created, allele frequencies were iterated over 50,000 generations by Equation A7-3 or until the stopping condition was satisfied. The stopping condition was $\dfrac{\sum_{i=1}^{i=k} |x_i' - x_i|}{k} < 10^{-12}$. At the end of this process, any allele with a frequency of less than 10^{-6} was considered lost.

CHAPTER 8

Heterogeneity-in-α Parameter Estimation

The observations consist of recorded deaths at times, $t_1, t_2, \ldots, t_k$. The deaths observed on day t_m, , are presumed to have occurred between times t_{m-1} and t_m.

If the initial number of adults in the cohort is N, then the observed mortality between times t_{m-1} and t_m is

$$\mu(t_m) = d_{t_m} / \left[N - \sum_{i=1}^{i=m-1} d_{t_i} \right].$$

The model estimate of mortality for the same time interval is

$$\hat{\mu}(t_m, A, \alpha, k) = 1 - \frac{p_{t_m}}{p_{t_{m-1}}},$$

where p_t is calculated from Equation 8-2. The least squares estimates are simply the values of A, α, and k that minimize the function

$$\sum_{i=1}^{i=k} \left\{ \frac{\left[\mu(t_i) - \hat{\mu}(t_i, A, \alpha, k) \right]^2}{\mathrm{Var}\left[\mu(t_i) \right]} \right\}.$$

The minimization was carried out with the *optim* R-function, which implements a Nelder-Mead (Nelder and Mead, 1965) procedure that doesn't require function gradients. Numerical integration of Equation 8-1 utilized the *distrEx-Integrate* R-function found in the *distrEx* R-package (Ruckdeschel et al. 2006).

Distribution of Age at Death

Each population was characterized by three parameters from the heterogeneity-in-α model, A, $\tilde{\alpha}$, and k (see Equation 8-1), and the sample size N. One sample consisted of N ages at death. For each of the N individuals in the sample, an age-dependent Gompertz parameter, $\xi\tilde{\alpha}$, was generated using the *rgamma* R function (with shape = k and scale = $1/k$) to generate the gamma random variable ξ. The age at death was generated using the inverse transform method (Fishman 1996), $\ln[1 - \xi\tilde{\alpha} \ln(1-U)/A]/\xi\tilde{\alpha}$, $U \sim$ uniform $(0,1)$. A total of 100 samples were generated for each *Drosophila* population. Only one sample was generated for the medfly populations.

Heterogeneity-in-α Model Fit to *Drosophila* Data

The data for the *heterogeneity*-in-α model fit to *Drosophila* are presented in Figures A8-1 to A8-4.

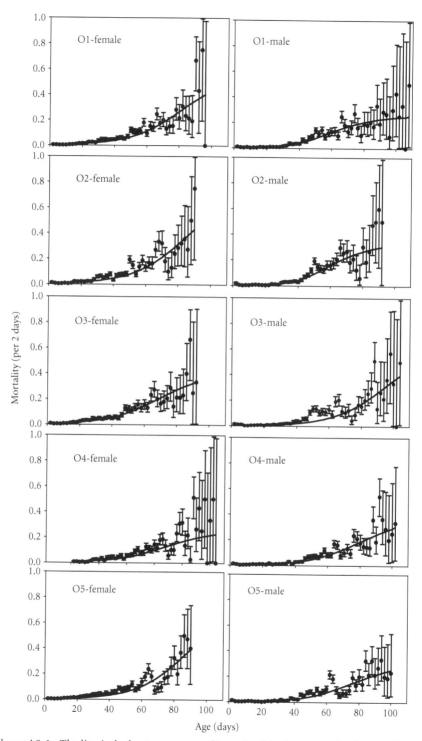

Figure A8-1. The line is the least-squares nonlinear fit of the heterogeneity-in-α model to the observed O population mortality. The circles show the observed 2-day mortality at each sampled age along with binomial 95% confidence limits.

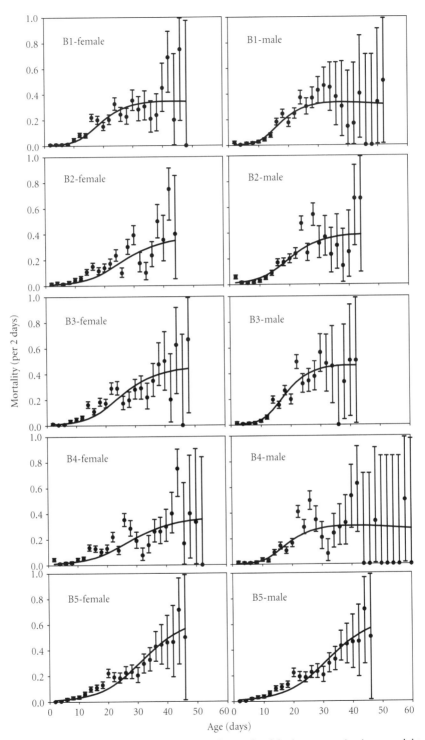

Figure A8-2. The line is the least-squares nonlinear fit of the heterogeneity-in-α model to the observed B population mortality. The circles show the observed 2-day mortality at each sampled age along with binomial 95% confidence limits.

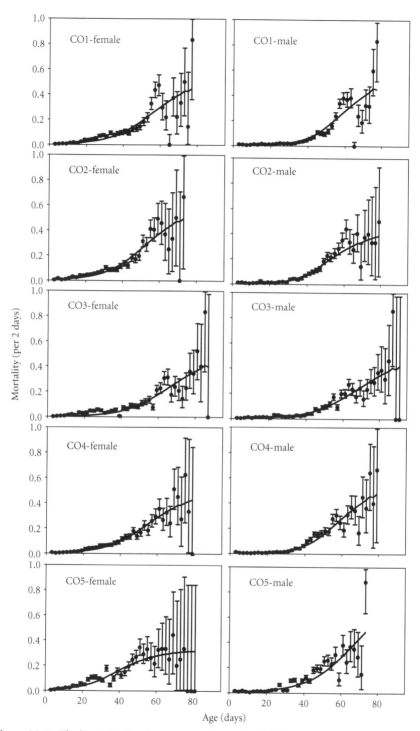

Figure A8-3. The line is the least-squares nonlinear fit of the heterogeneity-in-α model to the observed CO population mortality. The circles show the observed 2-day mortality at each sampled age along with binomial 95% confidence limits.

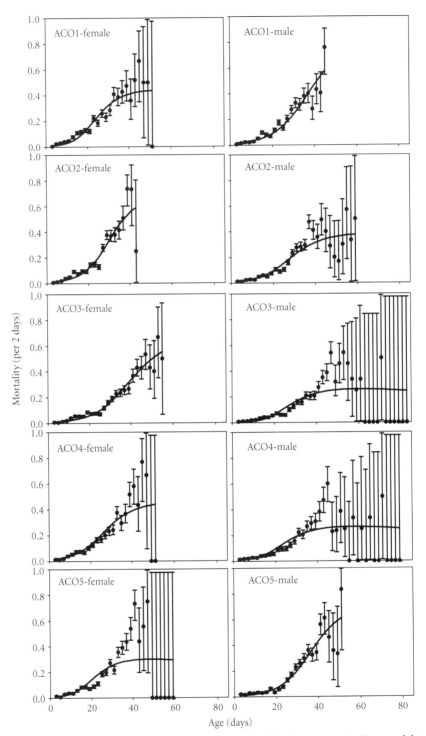

Figure A8-4. The line is the least-squares nonlinear fit of the heterogeneity-in-α model to the observed ACO population mortality. The circles show the observed 2-day mortality at each sampled age along with binomial 95% confidence limits.

CHAPTER 9

Application of the EHF Model to *Drosophila*

RESULTS FROM INDIVIDUAL FECUNDITY AND SURVIVAL RECORDS

The four-parameter EHF model (Equation 9-7) was fit to individual data as well as the five-, six-, and seven-parameter variants of Equation 9-7 (data from Rauser et al. 2005a). The success of the four models was then compared using the Akaike Information Criterion (AIC), the Bayesian Information Criterion (BIC), and a cross-validation index. We looked for the model that consistently had the smallest value of AIC, BIC, and the cross-validation index. To compute the cross-validation index, we divided the raw data in half. One half was used to estimate the model parameters. We then computed the mean predicted sum of squares with the second half of the data set. This process was repeated 100 times with different random partitions of the raw data. The average values of the cross-validation index are reported in Table A9-1.

Table A9-1. The model fitting results for three different data sets that included individual fecundity and survival records and four different EHF models. The lowest (best) value for each criterion is bold.

Model	Criteria	CO_{1-1}	CO_{1-2}	CO_{1-3}
4-par	AIC	**4.14**	4.55	5.17
	BIC	**4.67**	5.12	5.67
	Cross-validation	**3.93**	6.33	**9.92**
5-par	AIC	4.32	**4.28**	5.05
	BIC	4.98	**4.99**	5.68
	Cross-validation	5.01	**5.20**	11.48
6-par	AIC	4.34	4.37	4.83
	BIC	5.14	5.22	5.58
	Cross-validation	5.12	6.63	10.57
7-par	AIC	4.29	4.95	**4.33**
	BIC	5.23	5.94	**5.21**
	Cross-validation	4.59	6.10	10.23

RESULTS FROM INDIVIDUAL SURVIVAL RECORDS AND GROUP FECUNDITY RECORDS

Except for 1 case out of the 10 examined, the four-parameter model (Equation 9-7) had the smallest values of both AIC and BIC, as shown in Table A9-2. Accordingly, we focused on this model in our detailed analysis of the CO data.

Table A9-2. Summary of the stochastic fecundity model statistics. The four- ("4-par") and five-parameter ("5-par") models use fixed widths of 10 days. The lowest values of AIC and BIC are shown in boldface for each cohort.

Model	Parameter	CO1	CO2	CO3	CO4	CO5
4-par	AIC	**100.1**	**80.55**	**76.19**	63.72	**97.10**
	BIC	**100.8**	**81.16**	**76.64**	**64.20**	**97.85**
5-par	AIC	100.6	81.20	82.49	63.64	102.23
	BIC	101.5	81.96	83.21	64.25	103.17
6-par	AIC	100.1	81.02	76.78	**63.49**	98.38
	BIC	101.2	82.10	77.65	64.22	99.50
7-par	AIC	101.1	80.76	76.54	63.97	98.84
	BIC	102.4	81.83	77.55	64.82	100.15

REFERENCES

Abrams, Peter A., and Donald Ludwig. 1995. Optimality theory, Gompertz' law and the disposable soma theory of senescence. *Evolution* 49 (6): 1055–1066.

Aigaki, T., and S. Ohba. 1984. Individual analysis of age-associated changes in reproductive activity and lifespan of *Drosophila virilis*. *Experimental Gerontology* 19 (1): 13–23.

Bartke, Andrzej. 2005. Minireview: Role of the growth hormone/insulin-like growth factor system in mammalian aging. *Endocrinology* 146 (9): 3718–3723.

Baudisch, Annette. 2005. Hamilton's indicators of the force of selection. *Proceedings of the National Academy of Sciences of the United States of America* 102 (23): 8263–8268.

———. 2008. *Demographic research monographs: Inevitable aging? Contributions to evolutionary-demographic theory*. Berlin: Springer-Verlag.

Beard, Robert E. 1959. Note on some mathematical mortality models. In *The lifespan of animals*, Ciba Foundation Colloquium on Ageing, eds. G. E. W. Wolstenholme and M. O'Connor, 302–311. Boston: Little, Brown.

———. 1964. Some observations on stochastic processes with particular reference to mortality studies. *Transactions of the Seventeenth International Congress of Actuaries* 3: 463–477.

———. 1971. Some aspects of theories of mortality, cause of death analysis, forecasting and stochastic processes. In *Biological aspects of demography*, ed. W. Brass, 57–69. London: Taylor and Francis.

Bernstein, Adam M., Bradley J. Willcox, Hitoshi Tamaki, Nobuyoshi Kunishima, Makoto Suzuki, D. Craig Willcox, Ji-Suk Kristen Yoo, and Thomas T. Perls. 2004. First autopsy study of an Okinawan centenarian: Absence of many age-related diseases. *Journal of Gerontology Series A—Biological Sciences and Medical Sciences* 59 (11): 1195–1199.

Bohni, Ruth, Juan Riesgo-Escovar, Sean Oldham, Walter Brogiolo, Hugo Stocker, Bernard F. Andruss, Kathy Beckingham, and Ernst Hafen. 1999. Autonomous control of cell and organ size by CHICO, a *Drosophila* homolog of vertebrate IRS1-4. *Cell* 97 (7): 865–875.

Brent, Richard P. 1973. *Algorithms for minimization without derivatives*. Englewood Cliffs, NJ: Prentice-Hall.

Brooks, Anne, Gordon J. Lithgow, and Thomas E. Johnson. 1994. Mortality-rates in a genetically heterogeneous population of *Caenorhabditis elegans*. *Science* 263 (5147): 668–671.

Carey, James R. 1993. *Applied demography for biologists*. New York: Oxford University Press.

———. 2003. *Longevity: The biology and demography of life span*. Princeton, NJ: Princeton University Press.

Carey, James R., Pablo Liedo, Dina Orozco, and James W. Vaupel. 1992 Slowing of mortality rates at older ages in large medfly cohorts. *Science* 258 (5081): 457–461.

Carey, James R., James W. Curtsinger, and James W. Vaupel. 1993. Fruit fly aging and mortality. *Science* 260 (5114): 1567–1569.

Carey, James R., Pablo Liedo, and James W. Vaupel. 1995. Mortality dynamics of density in the Mediterranean fruit fly. *Experimental Gerontology* 30 (6): 605–629.

Carey, James R., Pablo Liedo, Hans-Georg Müller, Jane-Ling Wang, Damla Senturk, and Lawrence Harshman. 2005. Biodemography of a long-lived tephritid: Reproduction and longevity in a large cohort of female Mexican fruit flies, *Anastrepha ludens*. *Experimental Gerontology* 40 (10): 793–800.

Carnes, Bruce A., and S. Jay Olshansky. 2001. Heterogeneity and its biodemographic implications for longevity and mortality. *Experimental Gerontology* 36 (3): 419–430.

Chapman, Tracey, Jeffrey Hutchings, and Linda Partridge. 1993. No reduction in the cost of mating for *Drosophila melanogaster* females mating with spermless males. *Proceedings of the Royal Academy of London. Series B, Biological Sciences* 253 (1338): 211–217.

Chapman, Tracey, Sara Trevitt, and Linda Partridge L. 1994. Remating and male-derived nutrients in *Drosophila melanogaster*. *Journal of Evolutionary Biology* 7 (1): 51–69.

Chapman, Tracey, Lindsay F. Liddle, John M. Kalb, Mariana F. Wolfner, and Linda Partridge. 1995. Cost of mating in *Drosophila melanogaster* females is mediated by male accessory-gland products. *Nature* 373 (6511): 241–244.

Chapman, Tracey, and Linda Partridge. 1996. Female fitness in *Drosophila melanogaster*: An interaction between the effect of nutrition and of encounter rate with males. *Proceedings of the Royal Society of London. Series B—Biological Sciences* 263 (1371): 755–759.

Charlesworth, Brian. 1980. *Evolution in age-structured populations*. London: Cambridge University Press.

———. 1994. *Evolution in age-structured populations*, 2nd ed. London: Cambridge University Press.

———. 2001. Patterns of age-specific means and genetic variances of mortality rates predicted by the mutation-accumulation theory of ageing. *Journal of Theoretical Biology* 210 (1): 47–65.

Charlesworth, Brian, and Kimberly A. Hughes. 1996. Age-specific inbreeding depression and components of genetic variance in relation to the evolution of senescence. *Proceedings of the National Academy of Sciences of the United States of America* 93 (12): 6140–6145.

Charlesworth, Brian, and Linda Partridge. 1997 Ageing: Leveling of the grim reaper. *Current Biology* 7 (7): R440–R442.

Chen, C., J. Jack, and R. S. Garofalo. 1996. The *Drosophila* insulin receptor is required for normal growth. *Endocrinology* 137: 846–856.

Chippindale, Adam K., Armand M. Leroi, Sung B. Kim, and Michael R. Rose. 1993. Phenotypic plasticity and selection in *Drosophila* life-history evolution. I. Nutrition and the cost of reproduction. *Journal of Evolutionary Biology* 6 (2): 171–193.

Chippindale, Adam. K., Dat T. Hoang, Philip M. Service, and Michael R. Rose. 1994. The evolution of development in *Drosophila melanogaster* selected for postponed senescence. *Evolution* 48 (6): 1880–1899.

Chippindale, Adam K., Julie A. Alipaz, Hsiao-Wei Chen, and Michael R. Rose. 1997. Experimental evolution of accelerated development in *Drosophila*. 1. Larval development speed and survival. *Evolution* 51 (5): 1536–1551.

Christensen, Kaare, and James W. Vaupel. 1996. Determinants of longevity: Genetic, environmental and medical factors. *Journal of Internal Medicine* 240 (6): 333–334.

Christensen, Kaare, Matt McGue, Inge Petersen, Bernard Jeune, and James W. Vaupel. 2008. Exceptional longevity does not result in excessive levels of disability. *Proceedings of the National Academy of Sciences of the United States of America* 105 (36): 13274–13279.

Cichon, Mariusz. 2001. Diversity of age-specific reproductive rates may result from ageing and optimal resource allocation. *Journal of Evolutionary Biology* 14 (1): 180–185.

Clancy, David J., David Gems, Lawrence G. Harshman, Sean Oldham, Hugo Stocker, Ernst Hafen, Sally J. Leevers, and Linda Partridge. 2001. Extension of life-span by loss of CHICO, a *Drosophila* insulin receptor substrate protein. *Science* 292 (5514): 104–106.

Clark, Andrew G., and Rebecca N. Guadalupe. 1995. Probing the evolution of senescence in *Drosophila melanogaster* with P-element tagging. *Genetica* 96 (3): 225–234.

Comfort, Alexander. 1964. *Ageing: The biology of senescence.* London: Routledge and Kegan Paul.

Crow, James F. and Motoo Kimura. 1964. The theory of genetic loads. *Proceedings XIth International Congress Genetics* 2: 495–505.

Crow, James F., and Motoo Kimura. 1970. *An introduction to population genetics theory.* New York: Harper and Row.

Curtsinger, James W. 1995a. Density and age-specific mortality. *Genetica* 96 (3): 179–182.

———. 1995b. Density, mortality, and the narrow view. *Genetica* 96 (3): 187–189.

Curtsinger, James W., Hidenori H. Fukui, David R. Townsend, and James W. Vaupel. 1992. Demography of genotypes: Failure of the limited life span paradigm in *Drosophila melanogaster*. *Science* 258 (5081): 461–463.

Curtsinger, James W., and Aziz Khazaeli. 1997. A reconsideration of stress experiments and population heterogeneity. *Experimental Gerontology* 32 (6): 727–729.

David, J., J. Van Herrewege, and P. Fouillet. 1971. Quantitative under-feeding of *Drosophila*: Effects on adult longevity and fecundity. *Experimental Gerontology* 6 (3): 249–257.

Demetrius, Lloyd. 1997. Directionality principles in thermodynamics and evolution. *Proceedings of the National Academy of Sciences of the United States of America* 94 (8): 3491–3498.

———. 2001. Mortality plateaus and directionality theory. *Proceedings of the Royal Society of London, Series B—Biological Sciences* 268 (1480): 2029–2037.

de Grey, Aubrey D. N. J. 2007. Protagonistic pleiotropy: Why cancer may be the only pathogenic effect of accumulating nuclear mutations and epimutations in aging. *Mechanisms of Ageing and Development* 128 (7–8): 456–459.

de Grey, Aubrey, and Michael Rae. 2008. *Ending aging: The rejuvenation breakthroughs that could reverse human aging in our lifetime.* New York: St. Martin's Press.

Drapeau, Mark D., Erin K. Gass, Michael D. Simison, Laurence D. Mueller, and Michael R. Rose. 2000. Testing the heterogeneity theory of late-life mortality plateaus by using cohorts of *Drosophila melanogaster. Experimental Gerontology* 35 (1): 71–84.

Eaton, S. Boyd, and Melvin Konner. 1985. Paleolithic nutrition: A consideration of its nature and current implications. *New England Journal of Medicine* 312 (5): 283–289.

Ewens, Warren J. 1979. *Mathematical population genetics.* New York: Springer-Verlag.

Finch, Caleb E. 1990. *Longevity, senescence, and the genome.* Chicago: University of Chicago Press.

Fishman, George S. 1996. *Monte Carlo: Concepts, algorithms, and applications.* New York: Springer-Verlag.

Fleming, James E., Greg S. Spicer, Roger C. Garrison, and Michael R. Rose. 1993. Two dimensional protein electrophoretic analysis of postponed aging in *Drosophila. Genetica* 91 (1–3): 183–198.

Fontana, Luigi, Linda Partridge, and Valter D. Longo. 2010. Extending healthy life span—from yeast to humans. *Science* 328 (5976): 321–326.

Fowler, Kevin, and Linda Partridge. 1989. A cost of mating in female fruitflies. *Nature* 338 (6218): 760–761.

Fox, Charles W., and Jordi Moya-Laraño. 2003. Why organisms show late-life mortality plateaus: A null model for comparing patterns of mortality. *Evolutionary Ecology Research* 5 (7): 999–1009.

Fox, Charles W., Kristy L. Scheibly, William G. Wallin, Lisa J. Hitchcock, R. Craig Stillwell, and Benjamin P. Smith. 2006. The genetic architecture of life span and mortality rates: Gender and species differences in inbreeding load of two seed-feeding beetles. *Genetics* 174 (2): 763–773.

Frank, S. A. 2007. *Dynamics of cancer.* Princeton, NJ: Princeton University Press.

Fukui, H. Henry, Liang Xiu, and James W. Curtsinger. 1993. Slowing of age-specific mortality rates in *Drosophila melanogaster. Experimental Gerontology* 28 (6): 585–599.

Fukui, H. Henry, Lloyd Ackart, and James W. Curtsinger. 1996. Deceleration of age-specific mortality rates in chromosomal homozygotes and heterozygotes of *Drosophila melanogaster. Experimental Gerontology* 31 (4): 517–531.

Gadgil, Madhav, and William H. Bossert. 1970. Life historical consequences of natural selection. *The American Naturalist* 104 (935): 1–24.

Garland, Theodore, and Michael R. Rose, eds. 2009. *Experimental evolution.* Berkeley: University of California Press.

Gavrilov, Leonid A., and Natalia S. Gavrilova. 1991. *The biology of life span: A quantitative approach.* Chur, Switzerland: Harwood Academic Publishers.

———. 1993. Fruit fly aging and mortality. *Science* 260 (5114): 1565–1565.

———. 2001. The reliability theory of aging and longevity. *Journal of Theoretical Biology* 213 (4): 527–545.

Gerontology Research Group. 2010. *Verified supercentenarian cases—listed chronologically by death date.* Rev. April 28, 2007. Available at www.grg.org/Adams/I.HTM.

Geyer, Charles J., Stuart Wagenius, and Ruth G. Shaw. 2007. Aster models for life history analysis. *Biometrika* 94 (2): 415–426.

Gillespie, John H. 1973. Natural selection with varying selection coefficients—a haploid model. *Genetics Research* 21 (2): 115–120.

Gompertz, Benjamin. 1825. On the nature of the function expressive of the law of human mortality, and on a new mode of determining the value of life contingencies. *Philosophical Transactions of the Royal Society of London, Series B—Biological Sciences* 115 (1): 513–585.

Good, T. P., and Marc Tatar. 2001. Age-specific mortality and reproduction respond to adult dietary restriction in *Drosophila melanogaster*. *Journal of Insect Physiology* 47 (12): 1467–1473.

Gotthard, Karl, Sören Nylin, and Christer Wiklund. 2000. Mating opportunity and the evolution of sex-specific mortality rates in a butterfly. *Oecologia* 122 (1): 36–43.

Graves, Joseph L. Jr., and Laurence D. Mueller. 1993. Population density effects on longevity. *Genetica* 91 (1–3): 99–109.

———. 1995. Population density effects on longevity revisited. *Genetica* 96 (3): 183–186.

Greenwood, Major, and J. Oscar Irwin. 1939. Biostatistics of senility. *Human Biology* 11 (1): 1–23.

Haldane, John B. S. 1941. *New paths in genetics.* London: George Allen and Unwin.

Hamilton, William D. 1966. The moulding of senescence by natural selection. *Journal of Theoretical Biology* 12 (1): 12–45.

Hughes, Kimberly A. 1995. The evolutionary genetics of male life-history characters in *Drosophila melanogaster*. *Evolution* 49 (3): 521–537.

Ives, Philip T. 1970. Further genetic studies of the South Amherst population of *Drosophila melanogaster*. *Evolution* 24 (3): 507–518.

Jenkins, Nicole L., Gawain McColl, and Gordon J. Lithgow. 2004. Fitness cost of extended lifespan in *Caenorhabditis elegans*. *Proceedings of the Royal Society of London, Series B, Biological Sciences* 271 (1556): 2523–2526.

Johnson, Thomas E., Deqing Wu, Patricia Tedesco, Shale Dames, and James W. Vaupel. 2001. Age-specific demographic profiles of longevity mutants in *Caenorhabditis elegans* show segmental effects. *Journals of Gerontology Series A—Biological Sciences and Medical Sciences* 56 (8): B331–B339.

Joshi, Amitabh, Jason Shiotsugu, and Laurence D. Mueller. 1996. Phenotypic enhancement of longevity by environmental urea in *Drosophila melanogaster*. *Experimental Gerontology* 31 (4): 533–544.

Kannisto, Väinö. 1994. *Development of oldest-old mortality, 1950–1990: Evidence from 28 developed countries*. Odense Monographs on Population Aging No.1. Odense: Odense University Press.

Kannisto, Väinö, Jens Lauristen, A. Roger Thatcher, and James W. Vaupel. 1994. Reduction in mortality at advanced ages: Several decades of evidence from 27 countries. *Population and Development Review* 20 (4): 793–810.

Kenyon, Cynthia. 2005. The plasticity of aging: Insights from long-lived mutants. *Cell* 120 (4): 449–460.

Kenyon, Cynthia, Jean Chang, Erin Gensch, Adam Rudner, and Ramon Tabtiang. 1993. A *C. elegans* mutant that lives twice as long as wild type. *Nature* 366 (6454): 461–464.

Khazaeli, Aziz A., Liang Xiu, and James W. Curtsinger. 1995a. Effect of adult cohort density on age-specific mortality in *Drosophila melanogaster*. *Journal of Gerontology, Series A—Biological Sciences and Medical Sciences* 50 (5): 262–269.

———. 1995b. Stress experiments as a means of investigating age-specific mortality in *Drosophila melanogaster*. *Experimental Gerontology* 30 (2): 177–184.

———. 1996. Effect of density on age-specific mortality in *Drosophila*: A density supplementation experiment. *Genetica* 98 (1): 21–31.

Khazaeli, Aziz A., Scott D. Pletcher, and James W. Curtsinger. 1998. The fractionation experiment: Reducing heterogeneity to investigate age-specific mortality in *Drosophila*. *Mechanisms of Ageing and Development* 105 (3): 301–317.

Kingman, J. F. C. 1978. A simple model for the balance between selection and mutation. *J. Appied. Probability.* 15: 1–12.

Kosuda, Kazuhiko. 1985. The aging effect on male mating activity in *Drosophila melanogaster*. *Behavior Genetics* 15 (3): 297–303.

Kowald, Axel, and Thomas B. L. Kirkwood. 1993. Explaining fruit fly longevity (Technical Comment). *Science* 260 (5114): 1664–1665.

Leroi, Armand M., Adam K. Chippindale, and Michael R. Rose. 1994a. Long-term laboratory evolution of a genetic life-history trade-off in *Drosophila melanogaster*. 1. The role of genotype-by-environment interaction. *Evolution* 48 (4): 1244–1257.

———. 1994b. Long-term laboratory evolution of a genetic life-history trade-off in *Drosophila melanogaster*. 2. Stability of genetic correlations. *Evolution* 48 (4): 1258–1268.

Lindeberg, Staffan. 2010. *Food and western disease: Health and nutrition from an evolutionary perspective*. Chichester, United Kingdom: Wiley-Blackwell.

Loeb, Jaques, and J. H. Northrop. 1916. Is there a temperature coefficient for the duration of life? *Proceedings of the National Academy of Sciences of the United States of America* 2 (8): 456–457.

———. 1917. On the influence of food and temperature upon the duration of life. *Journal of Biological Chemistry* 32 (1): 103–121.

Luckinbill, Leo S., Robert Arking, Michael J. Clare, William C. Cirocco, and Steven A. Buck. 1984. Selection for delayed senescence in *Drosophila melanogaster*. *Evolution* 38 (5): 996–1003.

Mandel, S. P. H. 1959. The stability of a multiple allelic system. *Heredity* 13 (3): 289–302.

Marden, James H., Blanka Rogina, Kristi L. Montooth, and Stephen L. Helfand. 2003. Conditional tradeoffs between aging and organismal performance of *Indy* long-lived mutant flies. *Proceedings of the National Academy of Sciences of the United States of America* 100 (6): 3369–3373.

Matos, Margarida, Pedro Simões, Ana Duarte, Carla Rego, Teresa Avelar, and Michael R. Rose. 2004. Convergence to a novel environment: Comparative method versus experimental evolution. *Evolution* 58 (7): 1503–1510.

Maynard Smith, John, David J. P. Barker, Caleb E. Finch, Sharon L. R. Kardia, S. Boyd Eaton, Thomas B. L. Kirkwood, Ed LeGrand, Randolph M. Nesse, George C. Williams, and Linda Partridge. 1999. The evolution of non-infectious and degenerative disease. In *Evolution in health and disease*, ed. Stephen C. Stearns, 267–272. Oxford: Oxford University Press.

Medawar, Peter B. 1946. Old age and natural death. *Modern Quarterly* 1: 30–56.

————. 1952. *An unsolved problem of biology*. London: H. K. Lewis.

Miyo, Takahiro, and Brian Charlesworth. 2004. Age-specific mortality rates of reproducing and non-reproducing males of *Drosophila melanogaster*. *Philosophical Transactions of the Royal Society of London, Series B—Biological Sciences* 271 (1556): 2517–2522.

Mueller, Laurence D. 1987. Evolution of accelerated senescence in laboratory populations of *Drosophila*. *Proceedings of the National Academy of Sciences of the United States of America* 84 (7): 1974–1977.

Mueller, Laurence D., Theodore J. Nusbaum, and Michael R. Rose. 1995. The Gompertz equation as a predictive tool in demography. *Experimental Gerontology* 30 (6): 553–569.

Mueller, Laurence D., and Michael R. Rose. 1996. Evolutionary theory predicts late-life mortality plateaus. *Proceedings of the National Academy of Sciences of the United States of America* 93 (26): 15249–15253.

Mueller, Laurence D., and Amitabh Joshi 2000. *Stability in model populations. Monographs in Population Biology*. Princeton, NJ: Princeton University Press.

Mueller, Laurence D., Mark D. Drapeau, Curtis S. Adams, Christopher W. Hammerle, Kristy M. Doyal, Ali J. Jazayeri, Tuan Ly, Suhail A. Beguwala, Avi R. Mamidi, and Michael R. Rose. 2003. Statistical tests of demographic heterogeneity theories. *Experimental Gerontology* 38 (4): 373–386.

Mueller, Laurence D. and Michael R. Rose. 2004. Rules of evidence for models on trial (Letter to the Editor). *Experimental Gerontology* 39: 451–452.

Mueller, Laurence D., Casandra L. Rauser, and Michael R. Rose. 2007. An evolutionary heterogeneity model of late-life fecundity in *Drosophila*. *Biogerontology* 8 (2): 147–161.

Mueller, Laurence D., Parvin Shahrestani, and Casandra L. Rauser. 2009. Predicting death in female *Drosophila*. *Experimental Gerontology* 44 (12): 766–772.

Müller, Hans-Georg, James R. Carey, Deqing Wu, Pablo Liedo, and James W. Vaupel. 2001. Reproductive potential predicts longevity of female Mediterranean fruitflies. *Proceedings of the Royal Society of London, Series B—Biological Sciences* 268 (1466): 445–450.

Nagai, Jiro, C. Y. Lin, and M. P. Sabour. 1995. Lines of mice selected for reproductive longevity. *Growth Development and Aging* 59 (3): 79–91.

Nagylaki, Thomas. 1992. *Introduction to theoretical population genetics*. Berlin: Springer-Verlag.

Nelder, John A., and Roger Mead. 1965. A simplex algorithm for function minimization. *The Computer Journal* 7 (4): 308–313.

Novoseltsev, Vassili N., Janna A. Novoseltseva, and Anatoli I. Yashin. 2003. What does a fly's individual fecundity pattern look like? The dynamics of resource allocation in reproduction and ageing. *Mechanisms of Ageing and Development* 124 (5): 605–617.

Novoseltsev, Vassili N., James R. Carey, Janna A. Novoseltseva, Nikos T. Papadopoulos, S. Blay, and Anatoli I. Yashin. 2004. Systemic mechanisms of individual reproductive life history in female Medflies. *Mechanisms of Ageing and Development* 125 (1): 77–87.

Nusbaum, Theodore J., Joseph L. Graves, Laurence D. Mueller, and Michael R. Rose. 1993. Fruit fly aging and mortality. *Science* 260 (5114): 1567.

Nusbaum, Theodore J., Laurence D. Mueller, and Michael R. Rose. 1996. Evolutionary patterns among measures of aging. *Experimental Gerontology* 31 (4): 507–516.

Olshansky, S. Jay, Bruce A. Carnes, and Christine K. Cassel. 1993. Fruit fly aging and mortality. *Science* 260 (5114): 1565–1567.

Panter-Brick, Catherine, Robert H. Layton, and Peter Rowley-Conwy, eds. 2001. *Hunter-gatherers: An interdisciplinary perspective*. Cambridge: Cambridge University Press.

Papadopoulos, Nikos T., James R. Carey, Byron I. Katasoyannos, Nikos A. Kouloussis, Hans-Georg Müller, and Xueli Liu. 2002. Supine behaviour predicts time-to-death in male Mediterranean fruitflies (*Ceratitis capitata*). *Proceedings of the Royal Society of London, Series B—Biological Sciences* 269 (1501): 1633–1637.

Partridge, Linda. 1987. Is accelerated senescence a cost of reproduction? *Functional Ecology* 1 (4): 317–320.

Pearl, Raymond, John Rice Miner, and Sylvia L. Parker. 1927. Experimental studies on the duration of life. XL. Density of population and life duration in *Drosophila*. *The American Naturalist* 61 (675): 289–318.

Penna, Thadeu J. P. 1995. A bit-string model for biological aging. *Journal of Statistical Physics* 78 (5–6): 1629–1633.

Pinheiro, Jose C., and Douglas M. Bates. 2000. *Mixed-effects models in S and S-PLUS*. New York: Springer.

Platt, John R. 1966. Strong inference. *Science* 146: 347–353.

Pletcher, Scott D. 1999. Model fitting and hypothesis testing for age-specific mortality data. *Journal of Evolutionary Biology* 12 (3): 430–439.

Pletcher, Scott D., and James W. Curtsinger. 1998. Mortality plateaus and the evolution of senescence: Why are old-age mortality rates so low? *Evolution* 52 (2): 454–464.

———. 2000. The influence of environmentally induced heterogeneity on age-specific genetic variance for mortality rates. *Genetics Research* 75 (3): 321–329.

Pletcher, Scott D., and Claudia Neuhauser. 2000. Biological aging—criteria for modeling and a new mechanistic model. *International Journal of Modern Physics C, Physics and Computers* 11 (3): 525–546.

Popper, Karl R. 1959. *The logic of scientific discovery*. New York: Basic Books.

Prasad, N. G., Sutirth Dey, Mallikarjun Shakarad, and Amitabh Joshi. 2003. The evolution of population stability as a by-product of life-history evolution. *Proceedings of the Royal Society of London, Series B—Biological Sciences* 270 (1523): S84–S86.

Promislow, Daniel E. L., Marc Tatar, Aziz A. Khazaeli, and James W. Curtsinger. 1996. Age specific patterns of genetic variance in *Drosophila melanogaster*. I. Mortality. *Genetics* 143 (2): 839–848.

Promislow, Daniel E. L., Marc Tatar, Scott Pletcher, and James R. Carey. 1999. Below-threshold mortality: Implications for studies in evolution, ecology and demography. *Journal of Evolutionary Biology* 12 (2): 314–328.

Prowse, Nik, and Linda Partridge. 1996. The effects of reproduction on longevity and fertility in male *Drosophila melanogaster*. *Journal of Insect Physiology* 43 (6): 501–512.

Rauser, Casandra L., Laurence D. Mueller, and Michael R. Rose. 2003. Aging, fertility and immortality. *Experimental Gerontology* 38 (1–2): 27–33.

Rauser, Casandra L., Yasmine Abdel-Aal, Jonathan A. Sheih, Christine W. Suen, L. D. Mueller, and Michael R. Rose. 2005a. Lifelong heterogeneity in fecundity is insufficient to explain late-life fecundity plateaus in *Drosophila melanogaster*. *Experimental Gerontology* 40 (8–9): 660–670.

Rauser, Casandra L., Justin S. Hong, Michelle B. Cung, Kathy M. Pham, Laurence D. Mueller, and Michael R. Rose. 2005b. Testing whether male age or high nutrition causes the cessation of reproductive aging in female *Drosophila melanogaster* populations. *Rejuvenation Research* 8 (2): 86–95.

Rauser, Casandra L., Michael R. Rose, and Laurence D. Mueller. 2006a. The evolution of late life. *Ageing Research Reviews* 5 (1): 14–32.

Rauser, Casandra L., John J. Tierney, Sabrina M. Gunion, Gabriel M. Covarrubias, Laurence D. Mueller, and Michael R. Rose. 2006b. Evolution of late-life fecundity in *Drosophila melanogaster*. *Journal of Evolutionary Biology* 19 (1): 289–301.

Rauser, Casandra L., Laurence D. Mueller, Michael Travisano, and Michael R. Rose. 2009. Evolution of aging and late life. In *Experimental evolution: Concepts, methods, and applications of selection experiments*, eds. Michael R. Rose and Theodore Garland, Jr., 551–584. Berkeley: University of California Press.

Reed, David H., and Edwin H. Bryant. 2000. The evolution of senescence under curtailed life span in laboratory populations of *Musca domestica* (the housefly). *Heredity* 85 (2): 115–121.

Reznick, David, Leonard Nunney, and Alan Tessier. 2000. Big houses, big cars, superfleas and the costs of reproduction. *Trends in Ecology and Evolution* 15 (10): 421–425.

Rockstein, Morris, and Harry M. Lieberman. 1959. A life table for the common house fly, *Musca domestica*. *Gerontologia* 3 (1): 23–36.

Roff, Derek A. 1992. *The evolution of life histories: Theory and analysis*. New York: Chapman and Hall.

Rogina, Blanka, Tom Wolverton, Tyson G. Bross, Kun Chen, Hans-Georg Müller, and James R. Carey. 2007. Distinct biological epochs in the reproductive life of female *Drosophila melanogaster*. *Mechanisms of Ageing and Development* 128 (9): 477–485.

Rose, Michael R. 1982. Antagonistic pleiotropy, dominance, and genetic variation. *Heredity* 48 (1): 63–78.

———. 1984a. Genetic covariation in *Drosophila* life history: Untangling the data. *The American Naturalist* 123 (4): 565–569.

———. 1984b. Laboratory evolution of postponed senescence in *Drosophila melanogaster*. *Evolution* 38 (5): 1004–1010.

———. 1985. Life-history evolution with antagonistic pleiotropy and overlapping generations. *Theoretical Population Biology* 28 (3): 342–358.

———. 1991. *Evolutionary biology of aging.* New York: Oxford University Press.

———. 2007. End of the line. *Quarterly Review of Biology* 82 (4): 395–400.

———. 2009. Adaptation, aging, and genomic information. *Aging* 1: 444–50.

Rose, Michael R., and Brian Charlesworth. 1980. A test of evolutionary theories of senescence. *Nature* 287 (5778): 141–142.

———. 1981. Genetics of life-history in *Drosophila melanogaster*. II. Exploratory selection experiments. *Genetics* 97 (1): 187–196.

Rose, Michael R., and Laurence D. Mueller. 2000. Ageing and immortality. *Philosophical Transactions of the Royal Society of London, Series B—Biological Sciences* 355 (1403): 1657–1662.

Rose, Michael R., Linh N. Vu, Sung U. Park, and Joseph L. Graves Jr. 1992. Selection on stress resistance increases longevity in *Drosophila melanogaster*. *Experimental Gerontology* 27: 241–250.

Rose, Michael R., Mark D. Drapeau, Puya G. Yazdi, Kandarp H. Shah, Diana B. Moise, Rena R. Thakar, Casandra L. Rauser, and Laurence D. Mueller. 2002. Evolution of late-life mortality in *Drosophila melanogaster*. *Evolution* 56 (10): 1982–1991.

Rose, Michael R., Hardip B. Passananti, and Margarida Matos, eds. 2004. *Methuselah flies: A case study in the evolution of aging.* Singapore: World Scientific Publishing.

Rose, Michael R., Hardip B. Passananti, Adam K. Chippindale, John P. Phelan, Margarida Matos, Henrique Teotónio, and Laurence D. Mueller. 2005. The effects of evolution are local: Evidence from experimental evolution in *Drosophila*. *Integrative and Comparative Biology* 45 (3): 486–491.

Rose, Michael R., Casandra L. Rauser, Laurence D. Mueller, and Gregory Benford. 2006. A revolution for aging research. *Biogerontology* 7 (4): 269–277.

Rose, Michael R., Casandra L. Rauser, Gregory Benford, Margarida Matos, and Laurence D. Mueller. 2007. Hamilton's forces of natural selection after 40 years. *Evolution* 61 (6): 1265–1276.

Ruckdeschel, Peter, Matthias Kohl, Thomas Stabla, and Florian Camphausen. 2006. S4 classes for distributions. *R News* 6 (2): 2–6.

Service, Philip M. 1993. Laboratory evolution of longevity and reproductive fitness components in male fruit flies: Mating ability. *Evolution* 47 (2): 387–399.

———. 2000. Heterogeneity in individual morality risk and its importance for evolutionary studies of senescence. *The American Naturalist* 156 (1): 1–13.

———. 2004. Demographic heterogeneity explains age-specific patterns of genetic variance in mortality rates. *Experimental Gerontology* 39 (1): 25–30.

Service, Philip M., and Michael R. Rose. 1985. Genetic covariation among life history components: The effect of novel environments. *Evolution* 39: 943–945.

Service, Philip M., Edward W. Hutchinson, and Michael R. Rose. 1988. Multiple genetic mechanisms for the evolution of senescence in *Drosophila melanogaster*. *Evolution* 42 (4): 708–716.

Shahrestani, Parvin, Laurence D. Mueller, and Michael R. Rose. 2009. Does aging stop? *Current Aging Science* 2 (1): 3–11.

Shanley, Daryl P., and Thomas B. L. Kirkwood. 2000. Calorie restriction and aging: A life-history analysis. *Evolution* 54 (3): 740–750.

Shaw, Frank H., Daniel E. L. Promislow, Marc Tatar, Kimberly A. Hughes, and Charles J. Geyer. 1999. Toward reconciling inferences concerning genetic variation in senescence in *Drosophila melanogaster*. *Genetics* 152 (2): 553–566.

Shaw, Ruth G., Charles J. Geyer, Stuart Wagenius, Helen H. Hangelbroek, and Julie R. Etterson. 2008. Unifying life-history analyses for inference of fitness and population growth. *The American Naturalist* 172 (1): E35–E47.

Simmons, Michael J., Emily W. Sheldon, and James F. Crow. 1978. Heterozygous effects on fitness of EMS-treated chromosomes in *Drosophila melanogaster*. *Genetics* 88 (3): 575–590.

Spencer, Hamish G., and R. Williams Marks. 1992. The maintenance of single-locus polymorphism. IV. Models with mutation from existing alleles. *Genetics* 130 (1): 211–221.

Stearns, Steven C. 1992. *The evolution of life histories*. Oxford: Oxford University Press.

Steinsaltz, David. 2005. Re-evaluating a test of the heterogeneity explanation for mortality plateaus. *Experimental Gerontology* 40 (1–2): 101–113.

Steinsaltz, David, and Steven N. Evans. 2004. Markov mortality models: Implications of quasistationarity and varying initial conditions. *Theoretical Population Biology* 65 (4): 319–337.

Tatar, Marc, James R. Carey, and James W. Vaupel. 1993. Long-term cost of reproduction with and without accelerated senescence in *Callosobruchus maculates*: Analysis of age-specific mortality. *Evolution* 47 (5): 1302–1312.

Tatar, Marc, and James R. Carey. 1994a. Genetics of mortality in the bean beetle *Callosobruchus maculatus*. *Evolution* 48 (4): 1371–1376.

———. 1994b. Sex mortality differentials in the bean beetle—reframing the question. *The American Naturalist* 144 (1): 165–175.

———. 1995. Nutrition mediates reproductive trade-offs with age-specific mortality in the beetle *Callosobruchus maculatus*. *Ecology* 76 (7): 2066–2073.

Tatar, Marc, Susan A. Chien, and Nicholas Kiefer Priest. 2001. Negligible senescence during reproductive dormancy in *Drosophila melanogaster*. *The American Naturalist* 158 (3): 248–258.

Teotónio, Henrique, and Michael R. Rose. 2001. Perspective: Reverse evolution. *Evolution* 55 (4): 653–660.

Teotónio, Henrique, Margarida Matos, and Michael R. Rose. 2004. Quantitative genetics of functional characters in *Drosophila melanogaster* populations subjected to laboratory selection. *Journal of Genetics* 83 (3): 265–277.

Trevitt, Sara, Kevin Fowler, and Linda Partridge. 1988. An effect of egg-deposition on the subsequent fertility and remating frequency of female *Drosophila melanogaster*. *Journal of Insect Physiology* 34 (8): 821–828.

Tuljapurkar, Shripad. 1990. *Population dynamics in variable environments*. New York: Springer-Verlag.

Van Voorhies, Wayne A. 1992. Production of sperm reduces nematode life-span. *Nature* 360 (6403): 456–458.

———. 2002. The influence of metabolic rate on longevity in the nematode *Caenorhabditis elegans*. *Aging Cell* 1 (2): 91–101.

Van Voorhies, Wayne A., and Samuel Ward. 1999. Genetic and environmental conditions that increase longevity in *Caenorhabditis elegans* decrease metabolic rate. *Proceedings of the National Academy of Sciences of the United States of America* 96 (20): 11399–11403.

Vaupel, James W. 1988. Inherited frailty and longevity. *Demography* 25 (2): 277–287.

———. 1990. Relatives' risks: Frailty models of life history data. *Theoretical Population Biology* 37 (1): 220–234.

———. 1997. The remarkable improvements in survival at older ages. *Philosophical Transactions of the Royal Society of London, Series B—Biological Sciences* 352 (1363): 1799–1804.

Vaupel, James W., Kenneth G. Manton, and Eric Stallard. 1979. The impact of heterogeneity in individual frailty on the dynamics of mortality. *Demography* 16 (3): 439–454.

Vaupel, James W., and James R. Carey. 1993. Compositional interpretations of Medfly mortality. *Science* 260 (5114): 1666–1667.

Vaupel, James W., Thomas E. Johnson, and Gordon J. Lithgow. 1994. Rates of mortality in populations of *Caenorhabditis elegans* (Technical Comment). *Science* 266 (5186): 826.

Vaupel, James W., James R. Carey, Kaare Christensen, Thomas E. Johnson, Anatoli I. Yashin, Niels V. Holm, Ivan A. Iachine, Väinö Kannisto, Aziz A. Khazaeli, Pablo Liedo, Valter D. Longo, Yi Zeng, Kenneth G. Manton, and James W. Curtsinger. 1998. Biodemographic trajectories of longevity. *Science* 280 (5365): 855–860.

Wachter, Kenneth W. 1999. Evolutionary demographic models for mortality plateaus. *Proceedings of the National Academy of Sciences of the United States of America* 96 (18): 10544–10547.

Wattiaux, Jean M. 1968. Cumulative parental age effects in *Drosophila subobscura*. *Evolution* 22 (2): 406–421.

Waxman, David, and Sergey Gavrilets. 2005. 20 questions on adaptive dynamics. *Journal of Evolutionary Biology* 18 (5): 1139–1154.

Weindruch, Richard, and Roy L. Walford. 1988. *The retardation of aging and disease by dietary restriction*. Springfield, IL: Charles C. Thomas.

Weitz, Joshua S., and Hunter B. Fraser. 2001. Explaining mortality rate plateaus. *Proceedings of the National Academy of Sciences of the United States of America* 98 (26): 15383–15386.

Williams, George C. 1957. Pleiotropy, natural selection, and the evolution of senescence. *Evolution* 11 (4): 398–411.

————. 1966. Natural selection, the cost of reproduction and a refinement of Lack's principle. *The American Naturalist* 100 (916): 687–690.

Wilmoth, John R., L. J. Deegan, Hans Lundstrom, and Shiro Horiuchi. 2000. Increase of maximum life-span in Sweden, 1861–1999. *Science* 289 (5488): 2366–2368.

Wolfner, Mariana F. 1997. Tokens of love: Functions and regulation of *Drosophila* male accessory gland products. *Insect Biochemistry and Molecular Biology* 27 (3): 179–192.

Yashin, Anatoli I., Ivan A. Iachine, and Alexander S. Begun. 2000. Mortality modeling: A review. *Mathematical Population Studies* 8 (4): 305–332.

Yi, Zeng, Dudley L. Poston, Denese Ashbaugh Vlosky, and Danan Gu. 2008. *Healthy Longevity in China*. Springer Netherlands.

Young, Robert D., Louis Epstein, and L. Stephen Coles. 2009. Global mortality rates beyond age 110. *Rejuvenation Research* 12 (2): 159–160.

Zeng, Zhao-Bang, and C. Clark Cockerham. 1993. Mutation models and quantitative genetic variation. *Genetics* 133 (3): 729–736.

SUBJECT INDEX